BIRDGIRL

BIRDGIRL

LOOKING *to* THE SKIES IN SEARCH *of* A BETTER FUTURE

MYA-ROSE CRAIG

CELADON
BOOKS
NEW YORK

BIRDGIRL. Copyright © 2022 by Dr. MC Birdgirl Limited. All rights reserved.
Printed in the United States of America. For information, address
Celadon Books, a division of Macmillan Publishers,
120 Broadway, New York, NY 10271.

www.celadonbooks.com

Chapter-opening illustrations by Rhys Davies
Designed by Michelle McMillian

Library of Congress Cataloging-in-Publication Data

Names: Craig, Mya-Rose, author.
Title: Birdgirl : looking to the skies in search of a better future / Mya-Rose Craig.
Description: First U.S. edition. | New York : Celadon Books, 2023.
Identifiers: LCCN 2022036808 | ISBN 9781250807670 (hardcover) |
    ISBN 9781250807687 (ebook)
Subjects: LCSH: Craig, Mya-Rose—Childhood and youth. | Craig, Mya-Rose—Travel. |
    Bird watchers—Great Britain—Biography. | Environmentalists—Biography.
Classification: LCC QL677.5 .C773 2023 | DDC 598.072/34—dc23/eng/20220803
LC record available at https://lccn.loc.gov/2022036808

Our books may be purchased in bulk for promotional, educational, or business use.
Please contact your local bookseller or the Macmillan Corporate and Premium
Sales Department at 1-800-221-7945, extension 5442, or by email at
MacmillanSpecialMarkets@macmillan.com.

First published in 2022 by Jonathan Cape, part of the
Penguin Random House group of companies

First U.S. Edition: 2023

10   9   8   7   6   5   4   3   2   1

To Mum and Dad,
without whom none of this would have been possible

CONTENTS

BIRDGIRL

INTRODUCTION

I don't remember when I became obsessed with birds; it seems to me as though I've been birding forever. Given that my parents took me on my first twitch when I was nine days old, it's easy to see why I might feel that way. The towering bookshelves of our house are filled with titles such as *Shrikes*, *Sunbirds*, *Woodpeckers*, and *Nightjars*, beautifully illustrated guides to the birds of every corner of the world. As a child, and before I could even read, I would heave these books onto my lap to pore over their illustrations, tracing them with my fingertips, imagining I was stroking the soft feathers of a hummingbird, glimpsing a pitta in a shady rainforest, or basking in the reflected rays of the Regent Bowerbird's golden plumage. Later, I would copy the drawings into my own notebooks while planning extensive birding tours that would take my family all over the world.

By the time I was born, Mum, Dad, and my older sister, Ayesha, were already a well-known birding family, young and cool among the waterproof-clad, middle-aged, White male archetypes who made up the community at the time. Mum stood out from the crowd in other ways, too—she is Sylheti Bangladeshi, and the birding circle wasn't known for its ethnic diversity. We were "quirky" enough to be featured on a 2010 BBC documentary, *Twitchers: A Very British Obsession.*

Birds took me out into the British countryside and then beyond, to travel the seven continents with my parents, where I was exposed to not only rare, endemic, and magnificent new birds but also the impact of habitat degradation on people and wildlife. I witnessed biodiversity loss caused by climate change, logging, and palm oil plantations, as well as other overexploitation of the land.

Becoming a political and environmental activist felt like a natural progression, and one of my first steps into activism was inspired, unsurprisingly, by a bird: the tiny Spoon-billed Sandpiper, whose numbers were falling dramatically, owing to a loss of habitat on its breeding grounds in Siberia. Global warming was an issue, as was human destruction of mudflats in China and South Korea, where the birds break to feed during their migration south. Trapping, too, was a big problem for the Spoon-billed Sandpiper, a disastrous bycatch alongside the larger waders hunted for food by people scratching out a living on the birds' wintering grounds in Myanmar and southern Bangladesh. This was, very simply, a microcosm of the environmental issues facing the world.

In the summer of 2011, the global population of two hundred Spoon-billed Sandpipers weighed less than a single swan. Without human intervention, scientists believed it was highly likely that the species would become extinct within ten years. As part of a desperate effort to create a backup population—an ark, so to speak—conservationists transported thirteen young sandpipers from the Siberian tundra to Slimbridge Wetland Centre in Gloucestershire, only an hour from my home near Bristol. The captive population was added to the following year, when fourteen eggs were brought back from Russia and successfully hatched at Slimbridge. I remember hearing the news; it was an extraordinary and poignant moment, demonstrating what can be achieved when global conservation organizations work together.

The knowledge gained from creating this captive population led to a development called headstarting. Conservationists collected, incubated, and hatched the eggs on their breeding grounds and then hand-reared the chicks, readying them for their safe release. This process helped increase the number of fledglings that survive each year by 20 percent, and since 2015, 180 Spoon-billed Sandpipers have been released back into the wild. Today the population is estimated to be around one thousand birds.

In 2015, I was lucky enough to visit Sonadia Island in Bangladesh as part of an international task force to make a count of the migratory population. Every winter, the Spoon-billed Sandpiper departs its remote breeding grounds in far northeastern Russia to undertake an eight-thousand-kilometer migration, traveling along the coasts of Russia, China, and South Korea, down to Myanmar and Bangladesh. Just fourteen centimeters long, this sandpiper's comical spoon-shaped bill is very much part of its unique identity, and the tool it uses to sift through mud and silt in pools on beaches, mudflats, and other shallow wetland areas in its search for small invertebrates to feast upon.

As Mum and I boarded a motorboat to sail out to Sonadia Island's mudflats to begin counting, we had only one question on our minds: Were these winter visitors increasing or declining in numbers? It was a hot day, and a heat shimmer lay over the land. Was that a Spoon-billed Sandpiper in the distance? Yes! There were its weird bill, its fluffy white underbelly, and its speckled brown-and-gray wings. It was strange to finally see the bird in the flesh and feathers, knowing it had successfully made its epic trip.

This was a project I could get behind. The year before, I had launched my blog, *Birdgirl*, where I featured the many birds I had seen from around the world; now I added the Spoon-billed Sandpiper's plight to my pages, and in Dhaka I was able to use my growing social media platform to publicize the plight of the sandpiper through TV and national newspapers within Bangladesh as well as to the Bangladeshi diaspora in the United Kingdom. I was launched on a lifelong campaign to highlight the impact of climate breakdown and human destruction on our natural environment, on birds, on the land, and on people.

Birds serve as a kind of "canary in the coal mine" for climate change. The international effort for the Spoon-billed Sandpiper becomes all the more poignant when you imagine that the predicted one-meter rise in sea level would submerge not only Sonadia Island but also some 20 percent of Bangladesh—one of the most densely populated countries in the world—by 2050. A disaster for the sandpiper and humans alike. But . . . if we save the Spoon-billed Sandpiper, we will also save all the species of mammals, fish, and insect life that share its precarious habitat.

When I was seventeen, in 2020, I was invited to share a stage with the Swedish climate activist Greta Thunberg at the Bristol Youth Strike 4 Climate rally. I was a long way from the mudflats of Sonadia Island, and over the years, I had honed my message and my approach to activism. While the conservation projects of our birds and wildlife remain priorities for me, in front of an audience of forty thousand, I spoke about those without a voice: on behalf of Indigenous peoples removed from their ancestral lands in the name of conservation, and regarding the injustices visited upon the global south in the name of climate change action. I found my own voice as a young teenager, and while the journey feels much longer than the space of a few years, it is a voyage I intend to remain on.

Closer to home, and against the backdrop of my growing activism, family life was complicated. For much of my early childhood, Mum struggled with severe mental illness, veering between depression and mania, while my dad frantically searched for ways to make her feel better. Birds came to our rescue, time and again, a vehicle for healing, taking us out of our lives with bursts of fresh color, fresh wonder, providing enough sustenance to help us face whatever challenges lay in store.

I haven't enjoyed an easy migration into adulthood, but I hope to share, within these pages, how everything started with birds. There is nothing like the moment when your target bird appears. You may have been waiting for hours, staring into a bleak sky, chilled to the bone by the wind, or sweating in the stifling heat of a jungle, unable to swat away the annoying mosquitoes for fear of disturbing your bird. It's a moment of *Wow!* and *Look at that!* It will always be joyous, always a celebration.

And when you share these minutes with like-minded souls, there is nothing better—it's as if your soccer team has just scored the winning goal in a Cup Final. Backslapping, cheers, wide grins, and laughter ensue. And it's a feeling that lasts through the day, the next day, and beyond. The sight of a single bird blown off its migratory course, against all the odds, to spend a little time in a strange new land is a singular experience, a nirvana, a gorgeous creature to be burned into my memory forever.

1

My Family and Other Birds

GOLDEN PHEASANT

Native to the dense mountain forests of Western China, the Golden Pheasant has long been prized for its beauty, leading to its transportation around the world. Feral populations of escaped or released birds have established themselves across the globe—in the UK, US, Canada, Mexico, and various other European and South American countries, as well as in Australia and New Zealand. Records from as early as 1740 suggest this bird was the first species of pheasant brought to North America. Some historians have suggested that George Washington may even have kept a few at Mount Vernon.

My parents met in an underground club in Bristol called the Tube, an homage to the psychedelic 1960s. It was March 1995, and condensation dripped off the vaulted cellar walls as they made eye contact through a throng of gyrating bodies. The building vibrated to "Venus" by Shocking Blue, and the light show—a looped reel of original band footage—cast flickering images across their faces. In a secluded corner of the room, the banter began. My dad, Chris, introduced himself to Mum, Helena, as an electrician working in a chicken factory. She assumed he was the rough diamond among his university-educated friends. Only later did she discover he was referencing an obscure lyric from a song by the Specials, something he does to this day. He was twenty-seven years old with a background in hunt saboteuring, animal rights, and environmental activism; he was also, and most significantly, a birder.

When I ask Dad what drew him to Mum, he says, "Have you ever seen a photo of your mum at that age?" I have many pictures of Mum in her youth: dark-eyed and slim with very long, very straight black hair. There are also photos of my dad, often in a black turtleneck sweater and black jacket. His handsome face was framed by long, fair hair. It was not only his "look" that caught Mum's eye; she also loved his self-confidence and the way he held her eyes as he flirted.

While sparks flew at their first encounter, it hasn't been an easy road for my parents.

From her teens, my mother has struggled with mental illness, but she wouldn't be officially diagnosed with bipolar disorder until she was in her forties. Mum took her first overdose at fifteen, and by the time she went to university, she was acutely unwell. Flipping between manic and depressive episodes, Mum would have weeks when she went out clubbing every night followed by days in bed.

It was during one of Mum's manic episodes that she met her first husband; in the throes of another, she married him, in secret, in Sheffield.

Born into a fairly strict Muslim family, Mum, at first, had enjoyed her independence at university, but the combination of newfound freedom and mania was toxic. She began to party hard. My grandfather, unlike others in the Bangladeshi community, believed in education and sent all five of his children, including his three daughters, into higher education. He wanted Mum to succeed, but he also wanted her to marry a "suitable" man, and that meant an arranged marriage. Like many Muslim girls, Mum struggled with this clash of cultures. By the end of the first year, her grades were suffering, she was in love with an Englishman, and her mania had subsided into deep depression.

In her second year, she was convinced she would be chucked off her course and married off shortly after. Despairing, she took a handful of pills and fell into a deep sleep from which she could not be woken. Stomach pumped and still alive, Mum recovered enough to marry the Englishman in secret.

Of course, her parents went ballistic when they found out, but it was a done deal. She dropped out of her math and philosophy degree and began to consider the law as a career. When she found out she was pregnant with my sister, Ayesha, Mum enrolled at a local college, took and passed a couple of A levels (law and politics), and armed with good results—and a new baby—hand delivered her CV to fifty solicitors' firms, eventually landing a job as a paralegal. She then enrolled in a law degree. This snapshot is my mum in a nutshell: when she sets her mind to something, she will barrel on until she gets what she wants. Was she manic during this period? Definitely.

By the time she realized her marriage had been a mistake, she was well into her degree. She moved back to her parents' home, determined now that as a single mother she would have to support herself and Ayesha.

In 1994, she and my four-year-old sister moved out of my grandparents' house and across town to Clifton, a trendy part of Bristol. Working full-time, studying most evenings, and taking care of a toddler while still clubbing didn't faze her; she had manic reserves of energy, and her parents were still local enough to be relied on for childcare.

When she met Dad, she was still struggling, moving regularly from reckless party animal to lone wolf, mired in deep depression.

When Dad entered her life, she had little idea that he would introduce her to a world beyond the murky, sweaty interior of a 1960s nightclub. She was to meet creatures that would transform her relationship with nature.

But that was later; right now, in the underground club, Mum's friends had heard about this young man's strange passion for birdwatching. "Careful," they told her, "Chris is a *twitcher*." Of course, Mum had no idea what a twitcher was. (She thought it might have been slang for a drug addict, but in her eyes at least, it turned out to be something far worse.) While it didn't entirely break the stride of their relationship, from the outset Mum let Dad know he would be birding alone. "I'm Bangla. We don't do *birdwatching*!" she told him.

At first glance, you might imagine my parents' lives could not have been more different.

My Bangladeshi mother was born and bred in Bristol, and birdwatching, hill walking, hiking, rambling, and all other activities we associate with "the great outdoors" were not part of her childhood.

In 1955, my mother's father, my *nanabhai*, a Sylheti Bengali, immigrated to the United Kingdom when he was nineteen years old, without a penny in his pocket. He worked as a waiter at a thriving Indian restaurant in Oxford, the city awash with wealthy students. When he was ready to strike out on his own, he chose Bristol, full of equally well-to-do students and bereft of an Indian restaurant. Three years later, he opened the Taj Mahal, the first Indian eatery in the South West.

In 1961, he returned to East Bengal to marry my grandmother, my *nanu*. They had not met before the ceremony but came to the UK as husband and wife. Nanu arrived during one of the coldest winters Britain had seen in recent years, with deep snow on the ground, in a thin cotton sari and cardigan.

Theirs was a fairly traditional Muslim home by Western standards, but by South Asian standards it was liberal. My grandparents were passionate about education for their daughters, inspired by Indira Gandhi's attendance at Bristol's Badminton School in the 1930s. This hunger for education was unusual in Bengali families even as recently as the 1980s. My grandfather worked long shifts at the restaurant, rising at dawn to buy supplies from the fruit and vegetable markets. He put all five of his children through private school, understanding the value of academic success for the immigrant class. He was only too familiar with the experience of Bengali communities in East London, where sons left school at fifteen to work in the family restaurant and girls were packed off to Bangladesh in their teens to find suitable husbands. It wasn't an easy life for my grandparents; their ambitions were held in check by the hostile political landscape into which their children were born.

In the 1950s, racial discrimination was not illegal, and signs such as NO BLACKS, NO IRISH, NO DOGS could be seen in the windows of B&Bs, pubs, and flats to let. Nanabhai was no stranger to the taunts of racists who came into his restaurant at the weekend, looking for a fight. But he preferred rallies to protest against racism, rather than fighting with his fists on the streets. I have a photo of him at the 1963 Bristol Bus Boycott march. It's hard to believe that, during my parents' lifetime, a commercial business—in this instance, the Bristol Bus Company—could refuse to employ Black and Asian workers. The city united to boycott the company's buses, and the policy was abandoned. It was also widely considered to have influenced the passing of the Race Relations Act in 1965, outlawing racism in public places, and the act was extended a year later to prohibit employment and housing discrimination.

In the late 1970s, members of a far-right fascist party, the National Front, marched past my grandparents' house, chanting, "If you're White, you're all right." The family was terrified, the children instructed to keep away from the windows. Episodes like this and my *nanabhai*'s public fight against bigotry were the backdrop for my mother's and my aunts' burgeoning political awareness. From their early teens, they became vocal opponents of the oppression of South Africa's non-Whites by the policies of apartheid.

So, when Mum eventually qualified as a solicitor, she used her experience to highlight racial bias within her industry. She joined her law firm in 1996, one of only two non-White members of the staff. In the following years, she was instrumental in promoting ethnic diversity within the company. In the past, job applications had been discarded if the applicant hadn't achieved a first-class degree from a Russell Group university, or if they had a "funny name." The firm was awash with suited White men. Mum hired paralegals from all backgrounds, successfully proving to the partners that color or provenance of degree was no bar to excellence within the industry.

Dad was born in Rainhill, Merseyside, in 1968. His father worked in the laboratories of Imperial Chemical Industries, who were the big employers in the North West at the time. Academic and sporty, his dad was an all-rounder. My grandmother was from a working-class background, less confident in her academic ability but smart enough for grammar school. For a while, she worked for an insurance company, but like her peers, she gave it up to become a homemaker when she became pregnant. Dad's younger sister, Penny, loved horses and long walks with the family, while Dad was often happier in his own company.

From a young age, he was always more comfortable outside the house than in. They had a large garden with a six-foot fence, which he would scramble over to visit the woods nearby. He had a bird table and bird box in the garden, the latter remaining empty apart from one year when a male wren spent days building his domed nest inside, only for it to be rejected by the female.

On his seventh birthday he was given a book on how to identify European birds, and correctly named his first linnet in the next-door neighbors' front garden. His passion for birdwatching was born.

Family camping holidays in the Lake District, Wales, and South West England meant that the countryside was as much a part of his childhood and adolescence as going to political rallies was a part of Mum's.

When Dad was ten years old, the family moved to North Yorkshire, where his love for nature was fully realized. Now the family spent weekends walking the moors, climbing up to the impressive sandstone crags of the Wainstones or the distinctive half-coned peak of Roseberry Topping with its panoramic views across the Cleveland plain. Day trips to the picturesque Yorkshire coast between Runswick Bay and Staithes took in the fishing villages nestled in among the craggy cliffs and the coves and beaches. He hogged the binoculars and trained his eyes on the moors, the woodlands, the fields, fells, rivers, and lakes for birds, the pages of the guidebook burned firmly into his memory.

New birds came thick and fast. Here was the Red Grouse, ubiquitous amid the heather moorland, with its whirring wings and *go back, go back, go back* call as it burst into flight. It was hard to believe this plump-looking bird could achieve speeds of over a hundred kilometers per hour as it sought to evade an avian predator or, likelier, humans wielding guns. The Red Grouse is classed as a game bird, and around half a million are shot each year in the UK. While the moorland might appear to be natural, in many places it is intensively managed to increase the numbers of grouse. Other animals and birds that might eat the grouse are eradicated, and bird of prey are illegally persecuted in some areas.

Smart, golden-spangled, summer-plumaged Golden Plovers gave their fluting *peep* calls from hillocks among the heather.

Fulmar is a concatenation of the Old Norse words meaning "foul," as in foul-smelling, and "gull," which the Fulmar resembles in appearance. On the coast, they cruised off the cliffs on stiff wings, their innocent, wide-eyed stares disguising their ability to projectile vomit an odious-smelling liquid at anyone or anything coming too close to their nest sites among the noisy Kittiwakes, who sang their name from their coastal colonies perched on the sheer rock face.

Dad handwrote his first birding list at the age of ten, ninety-eight species spotted and noted! But it wasn't until a family holiday on Corfu that his ambitions ramped up. He was immediately seeing and identifying species he had only ever dreamed of: the red-eyed Sardinian Warbler, the cerulean-hued Blue Rock Thrush, and the Red-rumped Swallow nesting on the hotel balcony, appearing at first glance to be a cross between our house martin and swallow but, on closer inspection, clearly distinctive. It was a thrilling trip, a new and different experience for Dad; he was adding more birds to his repertoire, and his appetite for birdwatching expanded to fill every spare moment.

Before they met my dad, Mum and my sister had shown no interest in birding or even in the great outdoors. As a second-generation Bangladeshi growing up in Bristol, Mum had always considered herself a city girl. The closest she came to birding as a child was catching sight of the pigeons when she and her family went to the park to play cricket, or chasing away the gulls from her bag of chips on trips to the seaside. (Although now she admits she was rather enchanted by the tiny sparrows that flocked to her garden to be fed grains of leftover rice by my *nanu.*)

By the time Mum had actually worked out what dating a birdwatcher meant for their relationship, it was too late. She was in love. But she still wouldn't join him on the predawn expeditions to wherever the birders' pager alerts came from, where she imagined he would be standing for hours in the rain for the sake of seeing a single bird that had flown off its natural course. She wasn't wrong.

Now, on the weekends, Ayesha and Dad would disappear off to Chew Valley Lake or the nearby hills. Their very first trip out was a twitch to Cheddar Reservoir to see a Red Phalarope, a small, storm-blown gray-and-white wader that wears a little black mask. It breeds on the Arctic tundra, then spends the rest of the year out on the high seas, so when they occasionally, inadvertently, arrive on an inland reservoir, they are very tame, often never having encountered a human being before. Ayesha fell in love with this little bird spinning like a top on the surface of the water, sifting up the mud to dislodge the invertebrates below and feast.

Mum was starting to feel a bit left out. Both daughter and new boyfriend having abandoned her, she allowed herself to be drawn into their next adventure. "I'll go with you. Just once, to see what all the fuss is about."

The first family twitch was a cold day in March 1997, on a mission to Welney Wetland Centre in Norfolk in search of a Canvasback. This American diving duck had never been seen in the United Kingdom before. The alarm clock bleeped in the dead of night. (Just six months earlier, Mum would have been returning from a nightclub at this hour.) The game was on. Mum slept in the car all the way there and even nodded off in the bird hide. She was hardly slumming it, though—this hide had a carpet and plate glass windows; surely the most luxurious hide in the country, Dad had commented. What better place for her inaugural twitch?

Dad, Ayesha, and the other twitchers, who had by now gathered in this birding temple, trained their binoculars on the surrounding wetlands, a series of flooded fields with open flashes of water and part of the much larger Ouse Washes, a floodplain between the Old and

New Bedford Rivers. By now, Mum was growing impatient. This was *birdwatching?* Where was the bird? There was no sign of the Canvasback. They had all got up in the dark to drive halfway across the country, only to spend hours staring at a boggy swamp, and for *nothing.* Maybe the bird had decided to do the sensible thing and return to its home in North America, she wondered aloud.

But she was wrong; the Canvasback hove into view across the lake. Its distinctive long black bill and chestnut-brown head sat atop a pale gray body bookended by the black feathers on its breast and stern. A regal duck, escorted by a flock of its serene European cousins, the pochards.

My sister spotted it first and helpfully pointed it out to the assembled twitchers, who were just as desperate to catch a glimpse of this new visitor to our shores. She kept up a running commentary of its movements as it weaved its way through the pochard flock until, finally, everyone had seen it.

Mum was beginning to get it now, inspired by the excitement of the other twitchers, by Ayesha's exhilaration at being the first to spot the bird, by Dad's obvious pride.

Hours later, with the sighting of another new bird, Mum became fully initiated into this often misunderstood, often obsessive pastime.

On the way home, Dad decided he wanted to stop off to try to see a bird local to Norfolk: the Golden Pheasant. Although it originated in the mountains of Western China, the Golden Pheasant has an established population in the UK after being released into the wild by shooting estates. The Golden Pheasant is rare and extremely shy, often difficult to glimpse even for an experienced birder like Dad.

The sun was already starting to dip below the horizon when they arrived at Wayland Wood; Ayesha recalls the ground crackling with dry leaves, the air still and cold, a muntjac deer barking in the near distance. This foray wasn't so much about converting Mum to the art of birdwatching; it was about Dad's increasing desperation to find the Golden Pheasant. He had made this trip a dozen times in the past with no luck.

Leading the way, he scanned the undergrowth for a flash of gold, but Ayesha had begun to drag her feet, by now thoroughly exhausted by their predawn start. One circuit of the woodland was enough for her. And for Mum. The woods were eerily still. Leaves crunched and Dad was irritated, convinced the noise would scare off every creature in the area, including the reclusive pheasant. Eventually, in frustration, he left them to go and search on his own.

Mum and Ayesha waited for Dad to return. After a few minutes of near-total silence, the ground crackled behind them. Turning slowly, Mum laid eyes on a creature that took her breath away.

The Golden Pheasant is a study in primary colors: bright yellow feathers cover its back and head; its upper back is a deep iridescent green, with a fire-engine-red breast clashing against the blue of its wings. A riot of colors is topped off with a magnificent tail twice the bird's length. It strutted and scratched around in the leaves, foraging for berries and grubs, oblivious to the gawping faces of my mother and sister as they stared in wonder at this vision, now luminous in the light from the setting sun.

Mum had seen the pheasant *before* Dad, which made the taste of this rare encounter all the sweeter. Now she wanted to see *everything*. Mum claims it was this enchanted moment that made her a birdwatcher—and ignited the competitive streak my parents "enjoy" to this day.

She went on to twitch all over the UK, visiting the hot spots of the Shetlands, Fair Isle, and the Isles of Scilly in her enthusiasm to catch up with Dad. Sharing his love of birds with Mum catapulted Dad back to the early days of birdwatching, as he taught her everything he knew about identification and the importance of learning to wait patiently, whatever the weather.

Ayesha went on to become the youngest person to reach the incredible milestone of seeing four hundred species of bird in the UK when she was just twelve years old.

When Dad announced to his regular White middle-class family that he was bringing his new Bangladeshi girlfriend home for Christmas, *with* her six-year-old daughter, they didn't bat an eyelid.

It was different for Mum; she had led a dual life since her early teens, sneaking out of the house to go partying, dating whomever she felt like, regardless of their ethnicity. While my maternal grandparents weren't ultra-strict Muslims, clubbing and boyfriends weren't tolerated by the community—at least not in public. Even though Mum was a divorced single mother by the time she met Dad, she still couldn't openly walk down the street with him, in case they were seen together. Good Bangladeshi girls just didn't behave like that. The unspoken rule was less *Don't do that*, and more *Don't let anyone see you doing that*.

Ironically, Mum's first husband had also been a White non-Muslim boy, but she had eloped with him out of the blue, denying her parents the chance to object. Later, and in his favor, he had converted to Islam.

Dad, from the start, believed this secrecy was unhealthy; he couldn't square the need to lead two lives. He was also very hurt about being kept a secret; hadn't his own parents welcomed Mum and Ayesha into *his* family, just a couple of months after they started dating? He sort of understood Mum's fear of her family, but to a certain extent, Dad was culturally unaware; he had never been out with someone from a different culture or religion. When he finally met them, it felt like a job interview.

Dad was at a party with his mates, urged out of Mum's flat because her parents were coming over. Unbeknownst to her, they had heard rumors of Helena's new *English* boyfriend and decided it was time to meet him. When Dad arrived at Mum's flat, late and a little worse for wear, they were still there, patiently awaiting his return.

After answering formal questions about his education and employment, my *nanabhai* asked him the only question that really mattered: "Will you take care of my granddaughter?"

My sister was five years old when Dad came into her life. After their first official meeting, she declared to Mum, "I think he's good-looking apart from his body piercing. [He still has the nose stud.] He should be your boyfriend." Her own father wasn't around very much at the time, and Dad provided some structure.

Two weeks after they met him, my grandparents insisted Mum and Dad have a Muslim wedding; there was no way Mum was going to be allowed to parade a "boyfriend" around town without making a public commitment.

The wedding was to take place in Mum's flat in Clifton on a Sunday evening. As preparations were being made by her family, she and Dad were twitching a Little Bittern in Highbridge, South Somerset. This was a new bird for Mum and Ayesha, and Mum wasn't going to let her impending nuptials get in the way of a good twitch. Hiding in reed beds, they waited as the minutes ticked by. Eventually, Mum phoned home, pleading that the wedding be delayed for an hour; they couldn't leave without seeing the bittern.

It was a day of firsts for Dad: he had never even been to a Muslim wedding before, or met Mum's brothers until that night, and now, a nervous sweat soaking the collar of his shirt, he was reciting passages from the Koran at his own Muslim wedding. Of course, he had no Arabic and so had to repeat the complex sentences Mum's brothers threw at him, saying them again and again to get every inflection just right. Afterward, Mum dressed in a pure white salwar kameez, the family went out to eat. Their "official" wedding took place seven months later.

. . .

My family lives in an old miner's cottage on the northern slope of the Mendip Hills. It's incredible to think that 250 years ago, my bedroom would have maybe slept a family of ten, curled up across the uneven floorboards. Our house is on a dead-end lane that leads into dense woodland scented by wild garlic in the spring months. We can see rolling hills and a wide lake from our windows. In summer, wild roses and wisteria climb the walls.

I was raised in the Chew Valley, south of Bristol, where I enjoyed the freedom of the countryside and the fields of the Mendip Hills. I grew up watching birds fly to our feeders or flap their wings in the birdbaths, learning all the while how to distinguish between even the small brown varieties. In the warm afternoons, I would listen out for the call of the buzzards and, in the evenings, the ravens. In bed I would lay awake echoing the *hoohoo* of the Tawny Owl in the woods.

From an early age, I would roam the lane and the woods with my friends, exploring the animal paths and badger setts and the craggy old ocher mines. I would stick my hands into the wildlife pond, poking around for interesting creatures and removing them rapidly with a shriek when I found one. I followed the butterflies and bees about the garden, watched the moths flutter around our lights, and listened for the vague flap of bats' wings above my head in the dark. Every spring I lay down in woodland carpeted with bluebells; in the winter months I would toboggan the slope from the woods down to our house.

I grew up in a birdwatching family, devoting the weekends to checking which species had arrived on our patch or, farther afield, try-ing to catch whatever scarce bird had decided the cold shores of the UK were a suitable landing site. Driving from Somerset to Orkney and back over a couple of days was normal in my family. A long journey was never a deterrent, no matter how grim the weather. Twitching might sound full-on, but usually it's more a case of *We'll do what needs to be done.*

The TV series *Deadly 60* was the soundtrack to my childhood, featuring British naturalist Steve Backshall tracking down the world's deadliest predators, including the great white shark, the black mamba snake, and the polar bear. When I was ten years old, I went to see him live onstage in Bristol. He had a broken leg, which only made his mad adventures all the more compelling. Steve told his young audience that if we wanted to be TV nature adventurers like him, we had to study hard at school and take zoology degrees, but in the meantime, we should get out into nature, learning everything we could about wildlife and the outdoors. This resonated with me, and later I joined the Girl Guides *and* the Boy Scouts and embarked on activities such as climbing, abseiling, caving, potholing, filmmaking, wildlife sketching, camping, and making night shelters.

Meanwhile, I was pulling the bird guides off the shelves, soaking up the colors and habits of as many species as my brain could handle. At six years old, I read Gerald Durrell's *My Family and Other Animals*—a story about a chaotic family, deeply into nature—and saw my own life reflected.

Watching David Attenborough's documentary series *The Life of Birds* felt like a homecoming. Falcons, buzzards, hummingbirds, parrots, and terns filled the screen; this was not just a visual feast—it was a wish list. I was becoming enchanted by birds.

Imagine you're a passionate music lover: You have to have the music on wherever you are, in your living room, bedroom, or car. You play music to dance to, to relax to, or to cheer you up, depending on your mood. Music is so much a part of your life that you're unaware of the fact that you're "listening." That's a bit like loving birds for me. I don't so much notice them as absorb them—whether they're on my feeders in the garden, at the lake near my house, or on an Andean mountain ridge—and they always give me exactly what I need.

Birdwatching has never felt like a hobby; it's not a pastime I can pick up and put down but a thread running through the pattern of my life, so tightly woven that there's no way of pulling it free and leaving the rest of my life intact.

2

My Little Big Year

BLACK-BROWED ALBATROSS

Black-browed Albatrosses are monogamous, pairing for life, the fe-
male laying a single egg, which both she and the male incubate for
around seventy days. While young birds will return to the breeding
colonies beginning at three years old to practice mating rituals, with
75 percent of the world population found on South Georgia and the
Falkland Islands, they will only start to breed as of age seven. They
are long-lived, surviving up to seventy years in the wild.

By the time I was six years old, Mum and Dad both had busy jobs,
with Mum often working late into the evenings. Usually, I would
come home after school to the care of my sister, Ayesha, twelve years
my senior. For a long time, my favorite part of my day was the hours

between my sister picking me up from the bus stop and my parents returning home from work.

In front of the TV we'd eat Super Noodles, veggie nuggets, and chips while *Buffy the Vampire Slayer*, *Charmed*, and even *CSI* kept us entertained. I had to keep that last show a secret, naturally, which made it all the more exciting. At eighteen years of age, Ayesha was cool despite the fact she wasn't into birds so much anymore. It was as if she'd passed that baton to me, and I was running with it.

Years later, Mum, Dad, and I would attend family therapy sessions to deal with the impact of Mum's mental illness on the family, and it was recommended that before we start, my parents compile a list of major life events, including Mum's medical history. This document would save a lot of time in the sessions and would double as a "catch-up" sheet each time Mum was referred to a new psychiatrist. Mum, a seasoned birder, is a meticulous list maker. Everything is on the "Major Events: Chris & Helena" (MECH) spreadsheet, including an entry in December 2002: *Ayesha turns thirteen and becomes challenging.* My sister was a bit of a wild child, maybe not so much as Mum had been, but she certainly liked a party.

At just eighteen years old, she announced she was pregnant.

Dad was in America on business, a program manager with a multinational company, his hunt-saboteuring days behind him. His corporate transformation is not so dramatic as it sounds. He studied engineering at university and went to work for Hoover straight out of his master's. He quit after a couple of years, having saved enough money to go traveling. He let his hair grow long and believed this would be the pattern of his life: work, save, travel. Hair trimmed and back in a nine-to-five, he found himself enjoying his job and the promotions. When he met Mum, the roots he was already putting down went deeper.

His job often took him abroad, and he suspected Ayesha had waited for him to go away before she dropped her bombshell, scared of his reaction. He received an *Are you sitting down?* call to his hotel room in Washington, DC, from Mum and was on the next flight home.

Dad arrived back home on Saturday, and on Sunday, still shell-shocked from the news, he drove us all up to Norfolk to twitch a White-crowned Sparrow. He needed to "process," and birdwatching helped.

Dad was a "mindful" birdwatcher long before mindfulness had been "invented," or, at least, popularized. He enjoyed the complexity and focus of sifting through flocks of birds while he picked out different, often similar-looking, species. Concentrating on the slightest movement or noise giving away the presence of a half-hidden bird, counting and recording the number of birds, marveling at their antics and beauty drew him into the present, the here and now, where past and future concerns fell away. He says he achieves this without any "attempt" to achieve it. He needed to get into a better mental space to deal with Ayesha's news, and he knew that once he was out in nature, he would get to that space with zero conscious effort, which is, for Dad, a natural outcome of birding.

Back to the sparrow: the White-crowned Sparrow is commonly found in the United States, but it had appeared in the UK on only three previous occasions. And now it had turned up in Cley, not at the well-known bird reserve, a mecca for birders throughout the UK, but in the garden of a retired clergyman in the village nearby. The owners had very kindly put birdseed down on a long gravel path that ran alongside the house, and the bird would intermittently swoop from the beech hedge in which it was hiding to feast, all the while flashing its fancy white head stripes at the dowdy British sparrow cousins it was hanging out with. The throng of twitchers stood expectantly at the end of the driveway, waiting for its next performance.

The trip was a moment's respite from the fallout of Ayesha's news, from the stress and thousands of decisions that were yet to be made.

I *wasn't* shell-shocked but very excited by the prospect of having a niece or nephew. Full of bravado, I decided to share my news during a show-and-tell period at school. Usually, these sessions would feature one classmate or another bringing in an item from home and telling its story. Seashells and birthday gifts were presented, accompanied by tales of mini-breaks and birthday parties—one girl brought in her new kittens—but I had an announcement to make!

"My sister is pregnant!" I beamed. "I'm going to be an *auntie*!" My classmates cheered, but my teacher looked confused and a little alarmed. She knew Ayesha—my sister had attended the same primary school.

"Really?" she asked. "Are you sure?"

My teacher's skepticism made more sense later on, when our house became a battleground for arguments and tantrums. Ayesha was due to sit her A levels the following summer and then go to university, but instead, she moved out of the house a month before the baby was due, to live with her boyfriend. Mum and Dad didn't think much of the evasive answers both he and my sister gave to the questions they bombarded them with: *How will you support yourselves? How will you care for a baby? When will you take your exams?*

Neither of them was particularly keen to become a grandparent just weeks after their fortieth birthdays, but what choice did they have?

Before my sister moved out, she had been my second mother, waking me every morning, giving me breakfast before she walked me to the school bus, and meeting me at the same bus stop every evening; she would often be the one to put me to bed too. It's no wonder I started to feel twinges of jealousy toward this baby she was carrying. This baby who wouldn't have to compete with me for her attention, having already won. He or she would always come first from now on. Ayesha, I'm pleased to say, doesn't remember me ever showing a single glimmer of envy.

It was also a chaotic year in terms of birding. When Ayesha left home, our weekend forays were put on hold. There was no time to bird, to recharge; helping Ayesha prepare for this life-defining event and then helping her look after the baby became a priority. Although my sister, while she had cooled her birding heels to a certain extent, still enjoyed the odd family twitch, the reality of Ayesha's precarious future had wiped away some of the magic of birding.

At this time, Mum was full of nervous energy; looking after Ayesha while keeping up with her job was challenging, although I was oblivious to her struggles. Ayesha's situation filled every conversation and sapped my parents' strength, and by the end of the year, everyone, including me, was exhausted.

"I need to do something for myself," Dad announced. We were in the sitting room, watching TV, rain hammering against the cottage's dark windows.

"What thing?" I asked.

"A Big Year," he replied, looking at Mum.

"In January?" Mum said, looking puzzled. "Next year?"

"That's right." Dad's jaw was tight. Something odd was going on in this exchange, but I had no idea what.

"Laila's so small, though," Mum countered. "She's only four months old."

"I'm not going to the moon, Helena," Dad insisted. "It's just a Big Year."

This announcement was going to change the course of our lives.

A Big Year is a calendar year in which you try to see as many species of bird as possible within a given geographical area; in our case, this would be the UK. While Dad made it out not to be such a big deal, Big Years have been compared to extreme endurance sports.

The idea is to keep a "year list," where you record every species of bird you see from January to December, noting where you saw it, and when. Each bird is counted once and only once, and all are equal, the robin as valuable as a bird that has never been seen before in the UK.

Sighting three hundred or more species is a good round number for a year list in the UK without breaking any records. It's tough, but if you can go the distance, it's achievable, just like that endurance sport. The UK has unique peculiarities when it comes to year-listing. Our location means we see a disproportionately high number of vagrant species from different regions—Europe, Africa, Asia, and America, to name a few—compared with other countries. These are species that are not normally recorded in the UK and can turn up anywhere, anytime. There are, of course, patterns and good places to be at particular times of year, often at opposite ends of the country. You are trying to see all the species that regularly occur, some only in winter or summer or during migration in spring or autumn, some that appear only in specific areas of the UK. At the same time, you have to be responsive to news of rare birds turning up. Twitchers used to receive their bird alerts almost exclusively via pagers, and it is only relatively recently that pagers have been replaced by smartphone apps and social media such as Twitter. The messages from fellow twitchers and accounts such as Rare Bird Alert urge eager twitchers into action, tempting them with the glimpse of a rare bird.

Birds don't give a hoot about our Big Years and are often less than helpful; they fly off, hide when you're looking for them, and pop up in distant locations at inconvenient times. Sometimes they are identified only retrospectively from photographs on the internet. There is the British weather, too, of course, as well as the logistic challenges of reaching some of the more far-flung locations. All in all, it takes someone as obsessive as Dad to commit to twelve months of racing around the country in search of birds.

Dad made a year list every year—casually, not purposefully—a record of every bird that crossed his path, but in 2009, Dad wanted to see them all, and this would mean weekend trips away.

He explained he needed this project to feel "normal" again. He was forty years old, he insisted, not sixty; he wasn't ready to be a grandfather or full-time caregiver for Laila. It had been a tough year, and Dad had reached a breaking point.

"I'm doing it with or without you. But I *am* doing it," he told Mum. Meanwhile, I was trying, and failing, to imagine him going off birdwatching without us.

In December, he noted all the species he might encounter, from the common, such as blackbirds, robins, and starlings, to the more exotic or rare species, all the waifs and strays that wind up in the UK, blown off their course. His goal was the "gold standard": at least three hundred species. And it was an ambitious goal, one that would require him to spend most of his spare time birding.

Dad is someone who prepares carefully, makes sure that he has the right equipment, the terrain is safe, and he has enough warm clothing and flasks of tea for wet days. What he wasn't prepared for was my insistence that I go along, too, night or day, wherever he was called.

A whole year of weekends devoted to twitching? Tick.

Mum wasn't happy: I was only six years old. How would I cope with all the driving, the after-school twitches, the weekends without playdates? But I was just as determined as Dad. Mum *mostly* gave in. "You can do some of it, Mya, let's call it a *little* Big Year."

Mum admitted that, just like Dad, she, too, was exhausted and wanted to join us on the Big Year.

And Ayesha? While the Big Year sounds like a heavy commitment, the appearance of new birds ebbs and flows throughout the year; there would be many weekends and evenings when there was nothing to see, and Ayesha would have my parents' undivided attention. And while they had no intention of abandoning her, they were going to prioritize birding—for a while, at least.

The New Year's Eve party was still going strong when Dad left just after midnight to prepare for the first twitch of the year. Mum and I were going to miss it because neither of us fancied the early start. When Dad came home at lunchtime, with over fifty birds ticked off, Mum was also ticked off. It wouldn't take her too long to see all these birds once she started twitching, but that wasn't the point—her competitive streak had kicked in.

I began my *little* Big Year on the afternoon of January 1, 2009, a day of blue skies and blinding-white snow. We spent that first day birding within a few kilometers of home, noting the common species. I was having a great time stomping footprints into the snow and smashing frozen puddles while I breathed clouds of warm air into the cold wind. We made our way through farmland at Green Ore on the Mendips, scanning the fields and hedges, tracking down flocks of fieldfare, a more colorful version of the thrushes who join us in winter to escape the freezing conditions of their Scandinavian breeding grounds. A single Tree Sparrow and a Yellowhammer, once common farmland birds now pushed to the edges by modern farming methods, both put in an appearance.

We peered through the icy, twisting branches of woodland for Rooks, Chaffinches, and Sparrowhawks. We even lingered by Chew Valley Lake to make a note of the Mute Swans, Canada Geese, and Moorhens, followed by Blagdon Lake to find more geese. Canada Geese are native to North America but were introduced to the UK as long ago as the late seventeenth century, when they were added to James II's waterfowl collection in St. James's Park, London, and are now ever-present on the lakes of town and city parks as well as more rural water bodies throughout the country.

Although I had met all these birds many times before, it was as though I were seeing them for the first time because they were part of my Big Year. These familiar creatures and regular visitors to the local settings we had haunted for as long as I could remember felt suddenly as thrilling to look at as if they had been birds of paradise or Splendid Fairywrens.

I was adding dozens of birds to my list, but this was no mere box-ticking exercise. My list became a collection of unique experiences, my Big Year diary. That first day, as well as two dozen birds, I noted the weather: *Freezing frost. I am sitting in a field, it is a winter wonderland.* As my list grew, so did my drive to add more and more, regardless of how tired I felt, how cold the wind, or how much it snowed or rained.

Familiar birds were viewed through fresh eyes; I was more aware, more curious, conscious, and focused than I had ever been before as I identified and differentiated among species. Each sighting was a unique joy. The thrill and frustration of the chase only added to the momentum of the journey for the entire year. And along the way, my knowledge grew; I found that species I had always believed to be common apparently weren't so ubiquitous in January in Somerset. The Greenfinches I used to see in the garden on day one of the new year were now difficult to spot anywhere in the Chew Valley. I had to make a special effort to find the Smews that regularly visited Chew Valley Lake, but found that the Great White and Cattle Egrets could be spotted within a couple of kilometers of the house, birds that fifteen years ago my parents would have driven halfway across the country to catch up with.

When I was watching a very common bird—a Chaffinch, for example—I would think to myself, Imagine if this were a rare bird blown over from the Continent; we would be cooing over its stunning colors, but instead we barely notice them. This new perception was the singular most spectacular effect of the Big Year; I was appreciating everyday birds, watching each one as if it were a first for the UK, examining and noting everything about it, the tiniest details in the feathers, the nuance of color, admiring and feeling amazed each time I identified a new bird. This collection went into my diary, replete with drawings, pictures, details, and most of all, my love and respect.

My excitement quickly ramped up when Dad said we would be going after a Snowy Owl the next day. Could it be true? If it was, then this bird was way off course, as its natural habitat is above the Arctic tundra. If they do venture south, they usually fly no farther inland than northern Scotland, and even that is a rare occurrence. Apparently, the owl had been seen on the Isles of Scilly the previous autumn, and now, conveniently, it had made its way to the mainland, to the southern tip of Cornwall. We didn't even need to discuss it. Our target bird for January 2 was the Snowy Owl.

I have been twitching with my family since birth; some of the "lessons" of birding I had absorbed by osmosis, but everything else I learned from my dad, the expert. He was well known in the community before he met Mum, and with Ayesha and then me, our family became a fixture at big twitches. It might take years for a newcomer to absorb the birding "lore," the history and knowledge of what's gone before, but I had Dad to fast-track this process. Like all communities, the birding community has a culture. Imagine a child entering a museum or art gallery for the first time; she might race around the building shrieking until she learned that this environment requires a different standard of behavior. It's the same with birding: there's an etiquette. Learning how to wait patiently is key, and no one wants a moody kid who's tired or hungry expressing their frustration.

I learned stories of wild twitches and odd characters, their foibles and their triumphs. I was introduced to legends. "There are the four who traveled from Fair Isle to the Isles of Scilly in one day to get their bird," Dad might say, nudging me discreetly. Or, in awe, "There's the guy with the biggest list in the UK," and, "That guy, he's one of the fourteen who flew off Tresco in a helicopter to go and see a bird at Land's End."

Of course, there are practical skills too. I learned how to use binoculars correctly, how to avoid the big black patch that appears in the center of the lens the moment you put them to your eyes, how to focus, how to locate the bird within my binoculars, how to use a telescope and set up a tripod. I was "pishing" by the age of two. *Pish, pish, pish*, I would say, trying to coax a bird out of long grass and into the open. It's a strange sound that seems to inspire curiosity in birds, which will then, hopefully, emerge to investigate who is pishing at them.

More than that, I learned from my dad that birds are one part of a wider appreciation for the outdoors.

Meanwhile, the Snowy Owl beckoned.

Dad woke me up while the sky was still black. I wasn't new to dawn twitching and had prepared by going to bed in my thermals. I was lifted, still dozing, into the car. As soon as I felt the cold on my face, I woke up properly. Mum began to insist I try to go back to sleep, but I have never been able to do this on a twitch; I enjoy these early hours, watching a shadowy, slumbering landscape stream by.

We arrived just as the sky was turning a pale gray, parking up on the main road outside the village of Zennor in Cornwall, and headed inland, along a path that climbed up out of the valley and onto a hillside of scattered rocks and gorse interspersed with tall, dead, windblasted grass. It was very cold. As we approached, there was already a small gaggle of twitchers peering through their telescopes. What's the collective noun for a group of twitchers? I've heard it is an "anorak," and this group, us included, looked like an advert for people who don't know how to have fun, as we shivered in our parkas, hoping to catch sight of a single bird, which might or might not show up.

But there she was, a Snowy Owl, bigger, bulkier, and altogether more substantial than the male, standing at well over half a meter tall, perched on a hilltop, glowering at us with piercing yellow eyes, her razor-sharp talons gripping the ground. I couldn't help but feel that this fine bird looked out of place; she could have flown in straight from the set of a Harry Potter movie. Her pure white plumage, contrasting with heavy black bars, seemed to glow in the early-morning light. Puffy white feathers, reaching down to her feet, gave the appearance of snow boots, apt for her home in the Arctic but weirdly out of step in the grass, without a single flake of snow. Watching her take off, I could almost feel the broad wings vibrating overhead as the owl carved a path through the air. I drew a picture of her in my diary and wrote, *Owl was beautiful*. We could have spent hours watching her and, afterward, hours more wandering the Cornish countryside looking for other birds, but then a Mega Alert rippled through the pagers of the twitchers present: an alarm reserved only for the rarest of rare birds. Stunned silence: a bird none of us had seen before had been spotted in the North of England.

Returning home that night did not break the spell of our adventure. The cottage felt different, a pit stop on the way to somewhere more exciting. After a short rest, we set off again for the long drive to Teesside in North East England, where a Glaucous-winged Gull was apparently enjoying a break in the industrial wastelands of Cleveland.

Dad's mum, my gran Sylvia, lives in North Yorkshire, south of Teesside, so we paid her a visit. It was midnight by the time we arrived at her house, but she was more than used to our strange ways and reckoned any news of a rare bird close by would pretty much guarantee an impromptu visit. It was win-win for me; I got to see Gran and a rare bird.

Dad had spent his early teenage years birding in the area and felt right at home among the pipelines and semi-derelict industrial sites of declining heavy industry. Me, less so; it was quite a contrast to the beauty of southern Cornwall, but as Dad says, "Whatever it takes to get the bird!"

The bird had been frightened off earlier by a bird-scare gun. These guns, commonly used on farmland, are gas-powered devices fired off to scare away birds that have become a nuisance, and clearly the gulls, picking over the open garbage tip en masse, had become bothersome. There was no sign of the bird when we arrived, so we spent a couple of hours watching a flock of gulls picking over the detritus at the tip, waiting for the *right* gull to materialize.

And then it was back! Our target bird, the Glaucous-winged Gull, its bubble gum–pink legs popping in contrast to its monochrome plumage, flew in; we all let out a collective gasp—part joy, part relief—this was only the second time it had been seen in Britain.

Ayesha was missing us. She was struggling with a boyfriend who was struggling with becoming a father. He wanted to play football and go out drinking with his mates, and Ayesha didn't entirely blame him: he was only twenty-one; she often felt the same. Once or twice she confessed to feeling a little abandoned by us on the weekends. Mum and Dad persuaded Ayesha to go with us to Orkney later that year so we would be a birding family once more.

The next few months were spent in an idyllic frenzy as we traveled across what felt like every inch of the UK in our search for birds. I usually busied myself on long car journeys flicking through the well-thumbed and slightly torn family copy of the *Collins Bird Guide* to the birds of Britain and Europe.

In the spring we hiked halfway up Haweswater in the Lake District to see England's lone Golden Eagle. This was a place very special to

Dad: it was where, as a boy, he camped each year with his family and where he had seen the newly established pair of Golden Eagles, the only pair in England. Since his father's death, the Lakes were where Dad and I went to remember my granddad, visiting the special rock at Brothers Water where his ashes had been scattered into the lake. This was to be one of the last times we saw that female Golden Eagle—enormous but elusive, sitting tight on the high craggy cliffs she called home, all alone.

In late August we spent a week on the Isles of Scilly, camping in Hugh Town, the three of us packed into a tent that slept two and a half. We traveled over on the *Scillonian III* ferry from Penzance, bobbing like a cork on the sea, on a flat-bottomed boat without a steadying keel, which allowed it to enter the shallow waters around Scilly regardless of the tide. Watching the birds from the deck, we were rewarded with a Manx Shearwater and two Balearic Shearwaters.

On our first night of camping, we were disturbed by noises from the porch of our small tent, where our backpacks were stowed. A loud rustling and rummaging from just the other side of the thin fabric of the inner tent put Mum on high alert, her rodent phobia convincing her that a large rat was trying to come inside. If Dad felt the same, he kept it to himself. Muffled hissing and judiciously placed kicks from Dad against the tent were not discouraging the animal; finally, he tentatively unzipped the front flap. I am not sure who was more surprised, Dad or the enormous hedgehog. He leaped backward, and the hedgehog scuttled off into the night.

We had booked ourselves onto evening and daylong boat trips, on the lookout for seabirds—including the very rare Wilson's Storm Petrel, which frequents a small triangular area each summer out in the Atlantic, where it fishes. This was our only chance to see it. Majestic seabirds followed the boat, after the oily, stinky chum thrown out by the crew. On our second day out to sea, we saw two Wilson's Storm Petrels, tiny and fluttery as they were caught in the wind; after this, everything else was a bonus.

That year we spent many weekends on the drizzly Scottish islands—my diary noting it was *freezing, rainy*—and stopping anywhere and everywhere en route. We searched the Caledonian pine forests for the Crested Tit and Scottish Crossbill, the UK's only endemic bird. The male Capercaillie performing its mating dance and a pair of ospreys with their young from the RSPB (Royal Society for the Protection of Birds) hide at Loch Garten were the highlights.

My little Big Year was fun, and I was having the time of my young life.

One of the most exciting sightings of the year happened in July. We were back in Cornwall, sea-watching, my least favorite type of birding. The clue is in the name. Sea-watching is all about hanging around on headlands or cliffs for hours at a time, windswept and often very cold, waiting for something interesting to happen. Staring into gray skies and grayer water through telescopes or binoculars requires a special kind of concentration as you wait for an unusual bird to appear, and when something does appear, it's often very brief and no bigger than a speck on the lens. There is virtually no chance of identifying such specks. If you're twitching with others, often the shout will come up: "Look! Over there! Just by that wave—not that wave, the bigger one. Right of the white buoy…" It can be a frustrating experience, compounded by the fact that in less than a minute, the bird will probably disappear.

Bad weather, however, greatly increases the chances of seeing a good bird, given that strong winds blow seabirds toward the coast. So, while there is a greater opportunity of catching something interesting, you are also standing in a cold breeze and, often, the rain.

On the most southwesterly tip of the Cornish coast, Gwennap Head, we hiked up the coastal footpath, hoping a Cory's Shearwater would glide by on its bowed wings. Even though this would be an entirely new sighting for Mum and me, I was bored. Staring out to sea, watching heavy gray clouds wasn't my idea of birdwatching. I preferred to be on the move, in fields or woodland, on hills, by lakes and rivers—out of the wind and rain.

Dad, however, decided this was a "teaching moment": "Come on, Mya. Let's get realigned. Can you see the Runnel Stone?"

The Runnel Stone is a granite pinnacle in the sea one mile south of Gwennap Head. It used to be visible above water before it was struck by a steamship in 1923. Now a buoy marks its position, and it is commonly used as a directional aid among birders. Instructions such as "Third bird from the left at three o'clock of the Runnel Stone" are indispensable when the bird is no more than a dot in the sky.

Dad took my shoulders and positioned me. He knew how I felt about sea-watching, and the best way to reengage me was to teach me something. Locating gray birds against a gunmetal landscape is difficult, and it takes patience and practice; trying to spot the Cory's Shearwater among other birds would be hard. Over the years, I learned how to differentiate among species of seabird by the way they were flying. And the more familiar I became with identifying common birds, the better I could pick out an unusual one in their midst.

Eventually, I broke away from Dad and the other dozen or so twitchers and began to explore, climbing over the big rocks scattered across the cliffs. When I was bored of doing that, too, I made my way back to the twitching huddle and was about to tell my parents that I was hungry and maybe we should go back to the car for a flask of soup, when one of the party calmly stated, "Albatross." He was an experienced Cornish sea-watcher who, I was to learn, had scanned the coastlines of the local area for the past thirty years. Nothing much fazed him.

But...albatrosses are incredibly rare in the North Atlantic. They are almost mythical creatures in birding folklore, as they make improbable journeys from the southern oceans of the Antarctic. Flying nonstop for sixteen thousand kilometers is a feat worth celebrating, and if this twitcher had indeed spotted an albatross, surely he would be jumping up and down and waving his binoculars in the air!

"*There!*" he announced. After a few stunned seconds, everyone began scrambling for their telescopes. Giving directions to locations out at sea is notoriously difficult, usually a mess of "Over at eleven o'clock" or "By the boat! No, the other boat!" but thankfully this wasn't one of those occasions. I wasn't bored any longer as I watched, my mouth hanging open, a Black-browed Albatross cruising effortlessly toward us.

These beautiful birds have long, thin wings of over two meters in length that carry them vast distances across the oceans without a single flap as they ride the winds, an amazingly efficient way to travel. Imagine jumping off a Cornish cliff with only a hang glider for company and weeks later arriving in the southern oceans off Antarctica; that's the albatross. Soaring above the tumbling waves, slowly but surely it made its way toward our huddle of fourteen twitchers glued to their telescopes.

Each time I thought I had the best view possible, the Black-browed Albatross flew closer and I saw more, until it was only a few hundred meters away and I could see every detail. It flew close in, slowly toward Porthgwarra. I saw it with the telescope well and then with binoculars and then with just my eyes. It looped through the air by the cliffs and then swooped down toward the sea in a single elegant thrust, catching the wind beneath its massive wings. Rising again, it repeated the whole sequence twice more before flying parallel with the cliff face toward the next promontory—on which two lone birders were staring out to sea. We waved at them, jumped up and down, and screamed into the wind to try to catch their attention.

As the albatross melted away, the birders sprang into action, calling in the record, putting local birders on high alert, but this was to be the only sighting that day or in subsequent days, just sixteen of us in total to acknowledge it had ever happened.

When I uploaded the albatross sighting onto my online list, someone reported me for lying. Apparently there was no way a seven-year-old had seen an albatross out at sea.

By the end of August we had reached the magic figure of three hundred species in the UK with the sighting of a dainty little duck from North America, the Blue-winged Teal, in a pool near Portsmouth in Hampshire. This is the point where some year-listers falter, or "phase," as it is known in the birding world. Having reached the initial target, their appetite is sated and they lose the desire to carry on, but not us.

By September, I couldn't remember what I once did with my weekends before the Big Year or the birth of my niece, Laila. I would come home from school on a Friday, and Dad would tell Mum and me what time we had to be up in the morning. On this particular Sunday morning, we were in Carmarthenshire in Wales to look for a Glossy Ibis. As the year progresses, new birds are far less frequent, and the thin spread of birds in early September, before the migration season rush began, was a stark contrast to the earlier, busier months of our Big Year. The ibis twitch was especially tense because I had a friend's eighth birthday party to go to later that day.

We arrived at dawn on the outskirts of Burry Port, a small coastal town whose main claim to fame seemed to be the arrival of the aviator Amelia Earhart over eighty years before, making her the first woman to cross the Atlantic in a plane. (Her mysterious disappearance a few years later while attempting a circumnavigational flight around the globe is still unexplained.) In extremely thick fog, we scanned the marshy fields, but there was no sign of *any* bird. We would have to leave soon and I sensed my parents' agitation; they had promised me the party, but they badly wanted to see the Glossy Ibis.

The sun had finally risen high enough in the sky to burn away much of the fog, revealing two of the species. Almost black with hints of emerald green in their plumage, these large birds have rich brown underbellies. They stood in the middle of the field, feeding peacefully from the muddy waters with their long, curved bills. Nomadic by nature, these birds disperse from their breeding colonies in southern Spain as the long summer months dry out their habitat, influenced by the increasing impact of climate change. They head north, looking for new pastures, and Spain's loss is Britain's gain. While it's no longer classified as a rare bird, the Glossy Ibis isn't abundant on our shores, and any glimpse of it is an event.

As we raced back down the motorway to deliver me to my party, the call came in that another Glossy Ibis had turned up not three kilometers from our home in the Chew Valley at my local patch, Chew Valley Lake. Mum sighed; Dad and I rolled our eyes. This was the game, though—no rules, no timetable—as unpredictable as a vagrant bird.

At the end of September we headed off on what proved to be the longest journey of the year. A Sandhill Crane from North America had been sighted on Orkney off the most northerly tip of mainland Scotland. While I was at school and my parents were at work, the keenest twitchers had already made the pilgrimage north; we would have to wait until the weekend. It was only the third time the bird had been recorded in Britain and would be a new sighting for all three of us, but was such a long trip even possible over a weekend?

Mum and Dad decided it was. Around this time I was reading *Bill Oddie's Little Black Bird Book*, a hilarious peek into the birding community and the foibles and obsessions of the hard-core. I understood the inside jokes, and wasn't I on my own Big Year journey? If that didn't make me a part of the inner circle, then nothing would.

Plans were made; the bird was still there on the Friday morning, and by the time Mum and Dad got home from work, it was all systems go and we set off on the long journey north. Just in case the drive was not going to be difficult enough for Dad, we bundled Ayesha and thirteen-month-old Laila into the car as well. It was warm and cozy in the back seat, and we all settled down for a good night's sleep. Twelve hours later the car rolled up at a small quay near John o' Groat's, the most northerly point of mainland Scotland, with its wild and rocky coastline looking out over the dark seas of the Pentland Firth to our ultimate destination: Orkney.

Dad settled down for a much-needed power nap, but not for long. Laila, now wide awake and released from her car seat, decided that it would be fun to have a good crawl around the car, including over the prostrate body of her grandfather. All thoughts of dozing abandoned, Dad said we might as well look for some birds, and we were quickly rewarded with Black Guillemots and eiders in the small harbor.

Soon it was time to board the ferry for the ride over to Burwick on South Ronaldsay, the most southerly of the Orkney Islands. Dad kept a weather eye on the southern tip of the approaching island in case the crane decided to fly south just as we sailed north. Apparently it had already made a couple of experimental sorties over the waves on previous days but had thought better of it and returned to its favored fields. It seemed to take an age for the boat to dock and for us to disembark, anxiety rippling through every single twitcher aboard. The last bit of the journey is always the worst. The possibility of getting so close, having come so far, and then missing—or "dipping," as it is known by twitchers—the bird was too awful to contemplate. But, after a short drive on the other side, there stood the Sandhill Crane in the middle of a field of cut grass.

Almost a meter tall, this one needed no telescope or even binoculars to be singled out. Exhale and relax. Now we could all spend some time watching the long-legged, long-necked bird with its plumage of subtle grays and browns set off by a bright red forehead. The sun came out and everyone was satisfied. And then it was time to head back—we had a ferry to catch—but not before we spotted a nearby American Golden Plover, presumably caught up on the same winds that had brought over the Sandhill Crane. Two more year ticks for the list before Dad, exhausted, began the long journey south.

It should have been straightforward: a few hours' drive, find some accommodation, a good night's sleep for everyone, and then continue home. We called in at several places but everywhere was full, which was strange for Scotland in late September. After a few more failed attempts and another hour, Mum asked the latest fully booked hotel what was going on. Apparently we had arrived in the middle of a massive music festival. Dad groaned, but he had a plan.

My gran Sylvia had a static caravan on the northern edge of the Lake District; Dad not only had a plan, he had a key too. We arrived at three in the morning. The only thing we hadn't factored in was that Gran had decided to have an impromptu weekend away in her caravan. After recovering from the initial shock of being woken in the middle of the night, she was happy to see us all, baby Laila sealing the deal. We were windblown waifs and strays just like the birds we chased. The caravan became one huge bed, and I felt happy too. Ayesha and my tiny niece had been a part of a family birding adventure once more.

We arrived home late on the Sunday evening, and I headed off to school as normal the next day. I gave my standard answer to the usual Monday-morning question, "Do anything much this weekend?"

"Not really, Miss." I didn't know where to begin.

And so began the fracturing of my worlds: twitching with my family was one thing, and my life at school with my friends was another.

I wasn't aware of this gradual separation, but there was something vaguely overwhelming about the idea of having to explain to others why Mum, Dad, and I dropped everything at the bleep of a pager.

It needs to be said that, even for twitchers, our behavior was fairly extreme in 2009. From the sheer number of hours spent in a car to the way we would race off to catch sight of a bird at the bleep of a pager—regardless of responsibilities and commitments—it was a crazy and adrenaline-fueled year. Wild enough that it gained the attention of people outside the usual birdwatching community.

In the autumn, a TV production company wanted to feature my parents in their BBC Four documentary, *Twitchers: A Very British Obsession*. Mum and Dad were well known in the birding community, and when the producers contacted the Rare Bird Alert team for noted birders, the RBA told them about us. Our BUBO list (a live listing website where you record birds spotted) was consulted, showing that we were, so far, among the top five who had seen the most birds in the UK that year.

Dad's instinct was to turn them down. He didn't want to appear *on television* as a fanatical anorak. He had been in this business long enough to know how the media liked to paint these pictures. "They won't let nuanced details or even facts get in the way of a good story," he said. "And if I see one more 'Twitchers flock to see' headline...!" Besides, for him, the Big Year was pure escapism, and he really did not want anyone to get in the way of his birding.

When Dad was in his early twenties, there was a brief period when he lost himself to partying every weekend and long hours at work during the week. He went out on twitches, sure, but, for a while, birdwatching wasn't the all-consuming hobby of his youth. One Sunday morning, a little hungover but suddenly desperate to be outdoors, he went to the Cheddar Reservoir in Somerset. Sitting there without a backward or forward focus, just staring through the eyepiece of his telescope at birds, he felt his hangover melt away and was filled with a strong feeling of *I'm exactly where I need to be*. He realized this was what had been missing in his life, why he had felt so odd recently. He stayed there for hours, reclaiming an old enthusiasm and finding new energy for the outdoors. It was the first time he connected well-being with birding, and he has never looked back.

Dad's mental well-being was his main reason for embarking on the Big Year in the first place, and he argued that a TV program would make a "story" out of competitive birdwatchers while neglecting to mention the camaraderie and the mutual excitement of seeing a new bird.

"How am I supposed to enjoy myself with a bloody film crew at my back?" These were his final words, and Mum and Dad turned them down. But the producers were persistent, and they promised the show would focus on the autumn migration season as opposed to zeroing in on us as an obsessional twitching family. Finally my parents acquiesced.

A few weeks later, we began a nine-day shoot. Mum had cast the final vote, deciding it would be an amazing way to capture a special year, and while Dad had eventually given in, he still acted as if his teeth were being pulled out whenever any of the camera crew went near him.

It all started with a long interview with each of us. My parents were perfectly rational when talking to the cameras and did their best not to play into the eccentric birder stereotype. But the cameras, the interviews, and the attention blew me away; this was the most exciting thing that had ever happened to me.

At home on the first evening, I stayed up late, talking at length to the cameramen about why I was trying to see so many birds that year, when I should have just gone to bed. We were up before dawn the next morning, and I was sleepy and irritated when we arrived at the bird site, a very odd location for our first TV twitch. It was a working gravel pit in the West Midlands that, until the previous evening, had had an Aquatic Warbler residing in it. There was no sign of it when we arrived. I started moaning; I was tired, hungry, bored. Just usual seven-year-old antics, but the producers had caught the scent of a story. Now I was the kid who didn't really like birding but was forced into it by her overzealous parents. It made for interesting TV, I guess; they wanted a diva, I gave them one. I was still having a ball, exhausted or not.

By late autumn, there are fewer and fewer birds to see. Migration winds down and we were winding down, too, no longer up at dawn to chase around the country, bumping up our lists. But, as every twitcher knows, "the big one travels alone," which means that the rarest birds don't migrate with the other species. They turn up after everything else has already moved on. Blown in on errant winds, these are the birds you hope to catch sight of before the year is up.

It was a Thursday night in late October, and a wailing siren tore into the silence of our cottage. Dad had just put down the book he was reading to me and was about to switch off my light when we both jumped. It was an insistent tone, and it meant only one thing: a Mega Alert.

Earlier that day, a birder had posted a photo of what he thought was a Yellow-browed Warbler on a local online birding forum. It was a good record and the birder had every reason to be happy with himself for taking such an excellent picture. He had continued to chat online with a birding friend about the warbler photos, but then they had moved on to discussing the ID of an owl: Long-eared or Short-eared? All very interesting in a local context, but it wasn't going to set the birding world on fire. However, what happened next did.

The county bird recorder for the area was doing a final sweep of the local news sites to tally the day's sightings when he came across the photo of the "Yellow-browed Warbler." Dad read his post aloud: "This bird is actually an Eastern Crowned Warbler, a first for Britain." And then he added, "Oh. My. God."

The forum exploded with comments from hundreds of birders trying to get information about the bird and the location. The photographer responsible didn't even realize his "find" until he returned to his computer several hours later. The news was out, and twitchers from all over the country went wild.

Meanwhile, Dad was pulling his hair out. Mum wasn't at home and not answering her phone, and the film crew were calling, keen to wind up the program with one final family twitch. Dad was trying to follow the unfolding saga online while packing warm clothes and cold snacks for the three of us, as well as getting all our birding gear together so we would be ready to drive up to County Durham early the next morning. (Luckily Friday was an in-service day at school, so I had the day off.) When the film crew suggested he detour via London to pick them up so we could be filmed on the journey north, it was the final straw. "No way," he told them. "Get yourselves up there if you want to film us."

Early the next morning we piled into the car, ready to begin a five-hour journey. The sun was just starting to come up when we arrived at the site: an old quarry on the coast beside the town of South Shields. Despite the hour, the area was already teeming with twitchers, desperate to secure parking spaces as close to the site as possible.

The quarry was the perfect spot, forming a natural amphitheater around the stunted bushes and trees in which the warbler hid from view. We waited.

Hundreds of twitchers had gathered by now, calling in sick to work or taking a day's leave. There was one boy, older than me, who had clearly skived off school—and when the TV crew showed up, he pulled his hood over his face and retreated into the crowd. Smart.

Dawn broke; the minutes passed in silence, each one feeling like an hour, until the cry went up. There were twitchers poised above the quarry who had a clear view into the bushes. We shuffled and peered and pointed our binoculars. And then it appeared. The small green-and-yellow bird, which had flown to South Shields from East Asia, began to flit among the tallest branches of the trees. Its distinctive striped crown and sleek white belly identified it as the Eastern Crowned Warbler. It was a tiny dancer, performing turns in the air, playing to its mesmerized audience. It was the bird we had all been waiting for, and it seemed willing to reward us for our patience. It hovered in the trees and then dropped down into the bushes, adding to the drama. You could hear the sighs of relief and happiness from the enormous crowd as they gently jostled to get the best view through their telescopes whenever it appeared. One well-known twitcher lit a celebratory cigar. A euphoric twitch; everyone present had seen a rare bird really well.

"I saw it flying past!" I announced to the camera.

This is the great thing about the twitching community: the shared joy when a special bird is spotted by all. It's a very sociable group, and we regularly met up with familiar faces at birding hot spots, faces who went on to become friends. These days communication outside the twitch is far easier, given social media and dedicated birding forums.

There are different levels of twitcher. There are those who drop everything regardless of time and expense; these birders are hard-core and consist of around one hundred extremely dedicated individuals, but the less dedicated extend into the thousands. We weren't hard-core, and while some outside the birding community might have viewed us as obsessive, we didn't have the sort of lifestyle that allowed us to up sticks whenever an alert pinged on the pagers.

The little Eastern Crowned Warbler was the perfect bird, the perfect last big twitch of the year, and a coup for the documentary.

A year later, the reception for the BBC program was broadly positive with a side order of spite. The crew's original remit was to focus on autumn migration and the millions of birds leaving our shores to find a more hospitable southern habitat for the winter months. In reality, they had focused on the humans doing the watching, making the most of any competition between birdwatchers: who had seen what, when, and, most important, first.

When the documentary came out, it was to be my first brush with the toxic side of social media. Comments on popular birding forums concerned themselves with the welfare of the seven-year-old little girl who looked bored and unhappy. They were worried I was being denied a "normal" childhood because of my parents' insistence on dragging me up and down the country so they could follow their obsession. After all, they concluded, by the time I was ten, I would clearly be "into something else."

Mum wondered if the reaction would have been less critical had I been a boy. A boy would have reminded keen male birders of their own childhoods, after all. Maybe that's why it was so hard to accept that a young girl might genuinely be into birdwatching. This negativity stayed online, thank God. By now every twitcher in the UK had watched the program, and the reception from the birding community on the ground was friendly and warm.

The program had also caught the attention of my teachers, who insisted on sharing it with the whole class. I had to sit there, squirming, while my squeaky voice blasted from the TV set. My friends knew I loved birdwatching, but this was something else; there I was, on a fairly big screen, talking about it. I stole quick glances at my classmates, checking to see if they were as appalled as I was. Now they were laughing at the short clip of me asleep in the car. I was embarrassed; it was a new and uncomfortable feeling.

I saw 325 species in my seventh year and remain the only child in the world to have completed a Big Year. Of course, none of this would have been possible without my "obsessive" parents, but the Big Year had ignited my own hunger to see as many birds as I could, whatever it took.

Our victory and sense of accomplishment were bittersweet, however. By November, Mum was exhausted. She had given everything to the Big Year. And while Dad believed Mum's decline was a result of full-time work and full-time birding and being a grandma, it soon became obvious she was suffering from more than fatigue. The Big Year was over, and our lives could go back to normal, but none of us suspected that 2010 was going to be even more challenging.

{ 3 }

Shakira

SWORD-BILLED HUMMINGBIRD

Sword-billed Hummingbirds' bills are so long that, unlike other birds, they use their feet to preen their feathers instead of their bills. They have coevolved alongside a specific species of passionflower whose corolla tube matches the length of the hummingbird's bill. The flower depends exclusively on the Sword-billed Hummingbird to pollinate it, and in return the hummingbird is provided with high-quality nectar, which only it can easily access.

Instead of collapsing, Mum bounced back. By the time the new year rolled around, it was impossible to imagine she had been anything other than a busy, bustling, happy mum. She was first up in the mornings, breakfast on the table, all smiles and endearments before she climbed into her car and headed to the office. While in the past I had never been consciously aware of her mood flips—Mum was always just Mum whether she was on the hills with her binoculars, ecstatic, or lying in bed, staring sadly at the ceiling—I was now old enough to sense a shift in the house: Dad seemed happier when Mum wasn't tired, and this meant more birding at the weekends for all of us.

For a while she seemed unstoppable; immersed in her caseload, she was staying later and later at work, and in the evenings she would log back on to the computer to continue working. But the late nights and early mornings didn't leave much time for regeneration, and it was soon obvious that Mum wasn't sleeping. Dad knew this was always the first sign that she was on the verge of collapse. Sleepless nights led to panic attacks, which in turn fueled her frenzy. But Mum was good at letting me know she loved me, sneaking into my room if she got back very late and sometimes even waking me up with a small gift. We didn't have much time together, but I never felt neglected.

Her apparently limitless energy dried up in February 2010. While she was full of the joys one morning, the next she was suddenly lethargic, barely raising her head from her pillow to wish me a good day at school. When I was young, I wore my long hair in a single plait, and sometimes I would have to shake her awake to braid my hair. If I couldn't rouse her, I'd have to wear it in a fumbled, and deeply uncool, ponytail. If I noticed something was amiss with Mum, it was around tiny incidents like this.

I was eight years old; Dad had just picked me up from the bus stop, and Mum was in bed when we got home. By now it was early summer 2010, and Mum had just been signed off work with stress, anxiety, and depression.

As usual, I went to sit with her while Dad got dinner ready. There's one memory that sticks from this period: I'm watching Mum, who is lying in bed, her face turned to the window, while the sun bathes the room in golden light. How could she lie there, I thought, when there was so much to see, to do? Was she always going to be like this: energetic, happy, and essentially *Mum* one week or month or year, and despondent, unfocused, and miserable the next?

It made sense at first: we had just completed a Big Year, and Ayesha's pregnancy and new baby had taken it out of Mum, physically and mentally; of course she was tired. She kept saying that she needed a little rest and would be back at work in no time. It was confusing, but by now, it was also becoming *normal*.

The MECH list states that in July 2010, *H goes off sick from work. H starts taking antidepressant, citalopram 20 mg.*

It also notes, *Ayesha and Alex marry.*

In my memory, the wedding was fun. I got to hang out with my many cousins and watch my sister become a wife. Laila, almost two years old, was a happy little bridesmaid. Mum, meanwhile, hardly remembers the day. She was totally out of it on too much Valium, detached from everyone and everything. She held slurring conversations, barely in control of her words.

After this brief foray out of the house, she was in bed all the time, while still fully expecting to go back to the office this week or next, or the one after. Our household was under pressure in other ways besides: a couple of months earlier, Dad had been laid off. We had been living on Mum's income for several years while Dad's had paid off the mortgage; financially, we were okay—my parents were savers, and Dad's golden handshake and Mum's sick pay would keep us going.

Dad had been planning to "get out" for a few years anyway. His job as a program manager was, like Mum's, all-consuming, and he was keen to take some time off before trying out a different approach to work—a more flexible role, something to do with birds, nature, the environment.

Dad's break—the break he had been planning and looking forward to for so long—was put on hold as he was thrust into the role of full-time caregiver for both Mum and me. It was only years later that Dad confessed to me how resentful he had felt at the time: the "interlude" he had longed for had been snuffed out just like that.

Now he was around all the time, to see me to school and meet me off the bus in the afternoons. While I missed Ayesha, Dad was the only person who seemed to understand what was going on with Mum. When, finally, Dad confessed to me that Mum was *ill*, he reassured me in the same breath that she would get *better*. And I trusted him, and Mum. At eight years old, you have no wider frame of reference; if your parents tell you everything will be fine, why wouldn't you believe them? Mum would get better, return to work, join us on our birding expeditions, and generally just act more like a mum.

As well as making sure I had everything I needed, Dad also had to get Mum out of bed, ensure she was washed and dressed, that she ate *something* and didn't try to harm herself. And that took its toll on him. Very soon he was exhausted, and there was nothing I could do.

The Chew Valley in the summer is a haven for nature lovers. You only had to look out the kitchen window to feel the hills calling you. During this period, Dad and I would pull on our boots when I got home from school and head outside for an hour or so, to bird-watch or just hang out in the woods at the end of the lane. And when he felt he couldn't leave Mum on her own, we filled the feeders in the garden.

I reasoned that if she looked out her window, she might see robins, wrens, and even a charm of goldfinches. (While they are certainly charming, the collective noun for the goldfinch is actually derived from the Old English *c'irm*, which describes the birds' twittering song.) These beautiful birds, exuberant flashes of yellow and red, are frequent visitors to many garden feeders today, but that wasn't always the case. In the past, huge numbers of wild birds were caught and passed on to Victorian Britons who had a penchant for a caged bird. It's not always an advantage to be good-looking *and* a great singer! One of the earliest campaigns ever run by the RSPB was directed against this trade in wild birds. Maybe these tiny, perfect creatures would inspire Mum to join us.

But she wasn't getting any better, despite Dad's reassurances, and neither of my parents knew what to do; at the time, they were both hoping she was just burned out, that work had worn her down.

Our summer holiday beckoned, a trip to Ecuador that had been planned, booked, and paid for at the start of the year, when Mum was still working and feeling energized by the idea of a birding trip abroad. But, given that Mum was ill and Dad exhausted, was it still a good idea? they wondered.

Neither of them had a better one.

It was to be my first trip to South America, and while this hadn't been presented as a healing trip, Dad now began to consider whether it might not give Mum a positive focus, something she could grab hold of—anything to shift her attention away from herself and her depression for just a short while. It might be the push she needed to turn a corner. Dad felt that the Big Year had been a great success. Hadn't he spent every free minute of his life outdoors, in nature, with birds, urging Mum to do the same from the moment they met? When she had, he reasoned, it really worked for her, and maybe it would again.

"I'll try anything," said Mum. "Anything."

As if they didn't have enough to think about, they worried about me too. How would an eight-year-old cope with the tough birding itinerary Dad had so meticulously put together? This was to be no gentle ramble in the foothills of the Andes; we would be moving from lodge to lodge in and around northern Ecuador. All *I* cared about was hanging out with them with no interruptions while we looked for rare birds.

"The minute we start seeing the birds, all of this will feel very far away," Dad told Mum.

My parents understand each other better when they're birding; this was true then and it's true now. They share a special language that rarely requires one or the other of them to say, "Wow, look at that bird, isn't it amazing?" They just know. I guess Dad was hoping some of this magic would follow them into the rainforests of Ecuador.

Once it was decided the trip was definitely on, something shifted in Mum. We would be away for three weeks of nonstop birding; it felt like the start of the Big Year all over again. I was just glad Mum was out of bed, dressed, and happy to sit down with me and leaf through Robert Ridgely and Paul Greenfield's *Birds of Ecuador* to draw up our target lists. Dad was almost back to his old self as he browsed birding locations and memorized the rare and exotic birds he was most desperate to see; he was finally beginning to enjoy his "time off."

Mum always packs a stupid amount of stuff when we go traveling, prepared to meet any situation head-on. Thermals in a hot climate—tick; two hundred bird identification guides—tick again. But this time, we also had to pack a ton of books; if my parents were to "enjoy" this trip without the occasional whining from a bored eight-year-old cooped up in a car as we moved between locations, my suitcase would have to be at least half-filled with paperbacks. (And emergency Super Noodles: I was a very fussy eater.)

"This trip might be the best thing for us," Dad said optimistically as he loaded our bags into the car.

"Or it might be a disaster," said Mum with a sigh.

"Either way"—Dad heaved the boot shut—"we've packed enough shit to deal with it."

• • •

I didn't sleep for a single minute on the twelve-hour flight, unable to tear my eyes away from my guidebooks, poring over the tropical birds, imagining them perched high in the trees of the lush jungles I had seen on TV. There are around 1,600 species of bird in Ecuador, equivalent to 15 percent of the world's bird population. From the Andes to the Amazon, it is one of the principal destinations for world birders. With its abundance of rare and iconic birds, as well as its mountain ranges, rainforests, volcanoes, and the equator-straddling, warm seawaters, I felt as if I were about to enter paradise.

Pulling out the little notebook I had packed especially for this trip, I decided to make yet another list: "My Top Birds of Ecuador." For these, I promised myself, I would forgo food and sleep. While Mum and Dad dozed on either side of me, I selected three that sparked a strange feeling in my chest: excitement mixed with yearning. In the years to come, this feeling would come to define the moments of wonder whenever a strange new bird crossed my path.

The iridescent green wings of the Sword-billed Hummingbird seemed fantastical to me, a metallic shimmer that had no place in nature. Its comically long, rapierlike bill looked as though it had been attached as an afterthought, maybe to help it feed more easily. And then there was the enormous Harpy Eagle, one of the largest birds of prey, with razor-sharp talons and a full two-meter-wide wingspan. Its size alone, nearly a meter tall, gave me a thrill of fear, as did its eerie, pale face. And finally, the Andean Cock of the Rock, its head and chest plumage vibrant, so overwhelmingly *orange* that it swamped its facial features save for its two strange, staring eyes. These were birds of a different order. I underlined their names, added details, and, for good measure, attempted to draw them. If I ever lost my bird guide, I would have to identify these fantastical creatures from my sketches alone.

It was just past midday when we emerged from the Quito Airport, we were in Ecuador, and I was suddenly very tired. Mum had passed out the moment the plane took off and woke up only when we touched down. Dad had taken short, efficient naps, catching mealtimes and loo breaks. He was all business now, dismissive of my sudden weariness; this man had birds to see! He insisted the time difference would work in our favor in South America. That from now on, our body clocks would wake us up early—ideal for our dawn excursions. There was to be no nap for me, because we were heading straight into the mountains to begin birdwatching.

Organized bird tours are usually the best way to see lots of birds without spending a fortune or having to conjure up your own itinerary in a strange country. But we were going it alone, with just a local guide, Andrés, to lead the way. My parents hadn't been confident that Mum would be happy in a large group, and this felt like the best option. But Andrés's face had fallen when he first laid eyes on me at the airport. I was tired, complaining of the heat, and cross that neither of my parents was paying much attention to me: Dad hadn't looked up from his bird books, and Mum was busy chatting to Andrés. Our guide's last group had included a child of my age who had been bored and miserable the whole trip. Before we left the UK, Dad had convinced Andrés that I was just as obsessed with birds as any world birder. Now Andrés seemed equally convinced that Dad was wrong.

Ecuador is a little larger than the UK, and 20 percent of the country consists of national parks and reserves. For its size, Ecuador has the highest annual deforestation rate of any country in the Western Hemisphere. Experts believe that slowing the spread of deforestation and improving water management systems should be the country's national priorities. Today, Ecuador faces the expansion of large-scale mining operations in high-biodiversity areas with large numbers of endemic species and in Indigenous territories. The country's ongoing economic crisis and a dependence on fossil fuels will likely continue to incite clashes with communities who only want to protect their territories.

There are, however, many nature nongovernmental organizations (NGOs) working in Ecuador—for example, BirdLife International, Rainforest Trust's Conservation Action Fund, and FCAT Ecuador; the latter, composed of residents and scientists, is dedicated to the conservation of biodiversity in the tropical Andes, one of the world's most diverse yet threatened habitats on earth.

The Andes is a vast landscape. Everything is on a huge scale: mountains, forests, rivers, valleys, and skies. When you first plan to go to Ecuador—named for the equator, which "splits" the country in two—you don't think of cold days, colder evenings, freezing rain, or the wind. It's the altitude that makes the weather so extreme; there are also beautiful, clear, sunny days. You can almost taste the thin air at altitude, and sometimes your lungs may struggle for oxygen, a less pleasant sensation often accompanied by a low-key, humming headache.

The landscape also varies dramatically with altitude, from the barren, open mountaintops to the different types of forest as you head down the slopes and between specific altitudes. There is a sweet spot—occupied, for example, by Wildsumaco Lodge in the eastern foothills—at around 2,700 meters, where the number of bird species is at a maximum. You might see huge, mixed flocks of tanagers with various antwrens, spinetails, and even woodcreepers mixed in. Of course, different bird species fill every niche, so you "need" to visit the various heights and habitats. Heading down the slopes of the Andes, the weather is warmer, more humid.

Within hours of landing, we found ourselves winding up a mountain in our hired 4x4. Made up of an east slope and a west slope, the Andes in Ecuador are located around the central plateau of Quito. Many species of bird occur on one slope or the other. As well as their microclimate, the Andes are so high they present a natural barrier to many small species, in turn causing speciation, which is when bird populations evolve to become distinct species, on either slope.

Scatterings of snow drifted down from the clouds; the sky was full of brown birds. Where was the spectacular plumage of the South American creatures I had memorized from my guidebooks?

Tired and a little frustrated by the indistinguishable coffee-colored birds flying past our windows, I was also determined not to complain when we eventually pulled up in Papallacta, a village located on a barren mountain range at an altitude of 3,300 meters. From the air-conditioned car, the Sierra looked like a baked desert, but while the skies were a clear blue, the sun on full beam, it was cold enough for two jumpers.

Oblivious to the dust and rocks, Mum, Dad, and the guide scrambled out of the car and immediately began to identify the brown birds. There was no residue of Mum's fatigue and no sign either of the national bird of Ecuador: the Andean Condor. A type of New World vulture and yet another bird with an extraordinary wingspan, this was the only bird on my main list we had any chance of seeing that day. Hours later, in the late-afternoon sunshine, hungry and very dusty, we trudged back to the car with heavy legs, eyes dry and gritty from peering through a telescope for hours, and still no sign of the condor.

Back in the 4x4, we made our way to our lodge, arriving in the early evening, while it was still light. The air was full of squawking birds enjoying a last-minute feeding frenzy before they went off to roost, topping up with enough calories to carry them through the night.

Guango Lodge was well below the altitude of Papallacta, and in just a few hours, the landscape had transformed from the dry expanse of the Sierra to the rich forests surrounding our camp, although we were still high enough to suffer the background buzz of altitude sickness. The nights would be cold, requiring full thermal coverage. In the midst of these wooded mountains, our lodge was rustic, with wooden cabins and a central dining hall surrounded by colorful hummingbird feeders where, in the days to come, we would eat hot soup and drink hot chocolate, both staples in the Ecuadorean Andes.

I was longing for my bed, desperate for a soft pillow and to shut my eyes for at least eight hours—or until dawn, whichever came later. But Mum insisted I eat first. I dragged my feet as she led me toward the dining hall. I hadn't seen a single bird on my list, and at that moment, I was too exhausted to care. I opened my mouth to ask her to slow down, to wait for me, when a burst of color in motion caught my eye.

By this point, you might think I was a fully committed birdwatcher. I had just completed my first Big Year, and I had been birding since I . was nine days old, but I remember the moment when birds became the absolute center of my world, and that moment was right now, on the first day of our trip, when I was at my least observant.

It wasn't the bright red and green feeders, which hung from the branches of the nearby trees, that had stopped me in my tracks but the hummingbirds, darting frantically between them in their perpetual hunt for more food. I stopped breathing.

In the low sun, they sparkled in shades of luminous turquoise, emerald green, and deep velvety violet. I once again puzzled over how such colors could exist in nature. And their wings! They flapped at such speeds they were barely visible, emitting a soft whirring sound; I thought of bees hovering over a field of daisies in summer. No picture in a guidebook could touch their extraordinary depth of color, their velocity, or their charm. How could a photograph come close to capturing the ethereal quality of these tiny creatures, which seemed to exist in another dimension, oblivious, in their frenetic buzz of activity, to this world?

The hummingbirds didn't seem to care that Andrés had appeared in their midst and was gently stroking the soft feathers on their backs as they drank from the birdbaths, or that my parents were attempting to do the same. I forgot about my fatigue and supper and stared at the birds until I was dragged off to bed, their colorful bodies filling my dreams.

The next morning I made straight for the feeders, eager to start identifying the hummingbirds. Sipping hot chocolate and making notes, I was in heaven. I didn't care about the rest of Ecuador—I had everything I needed right here. As I watched their feverish bodies streaking bright flashes in the sunshine, a new and different bolt of color suddenly caught my eye, and I moved closer. A tiny bird was moving in a single blur of iridescent green. Its feathers gleamed in the sparks of light striking through the canopy of leaves overhead. The Sword-billed Hummingbird was putting on a show *just for me.*

How could this bird even fly? Its impossibly thin bill was longer than its body, yet it took to the air gracefully, its wings whirring like a clockwork toy. It was a hummingbird like all the others, but it was in a class of its own. It needed an entirely different posture from all the other birds just to support its long bill, but at the same time, it was elegant, balancing itself in the air with as much dignity as its cousins.

It was in this moment, as I watched it defend its territory from the other species, that I decided I *adored* hummingbirds.

Eventually, it flew away and I was left nursing a cold drink, but I didn't move; an idea was brewing. Later, I announced to Mum and Dad that I would see every one of the 374 species of hummingbird in the world. It was a bold statement for an eight-year-old, and it came to embrace not only hummingbirds but all species of bird, wherever and whenever. My parents responded enthusiastically, and I began a birding journey that hasn't faltered.

Rested, fed, and inspired, we had set the pace for the trip. My appetite for new birds was huge, and with each I was spurred on to work harder to see more. It didn't matter that I was eight years old; my stamina easily matched my parents'—at times, it exceeded theirs.

Andrés stopped looking cross whenever I opened my mouth and, once or twice, even laughed at my stupid jokes.

Entering the forest with Mum and Dad and Andrés on our second day had been a daunting exercise. A staggering number of birds gleefully flew in and out of the trees; I was immediately overwhelmed. How would I ever begin to identify a single bird in here? There were too many, and they were so frenetic. I couldn't focus on one before another took its place in the trees or on the ground or in the air.

"It's okay, Mya," began Dad, dragging himself away from the cacophony overhead. "You're going to learn every bird by the time we leave, but let's start looking at the most common ones first. Who's that guy?"

Dad was right; I just needed to take it slower. "Tanager?" I said. Dad nodded. I counted twelve tanagers.

"Now let's move on to this chap." And in this way, I was able to identify a mass of birds, slowly, methodically, naming them, placing them into a species group, and moving on to the next.

In the days that followed, I refused to sleep during our siesta time because the idea of napping while the sun was up made no sense. Instead, I would spend these hours reading, losing myself in *Watership Down* and books by Jacqueline Wilson or Michael Morpurgo, in which animals, not humans, were the heroes of the day. If I wasn't reading, I was hovering by the hummingbird feeders, and sometimes, I would try to stroke the birds, just as Andrés had done, but they darted away from my fingers every single time, disgusted by the cheek of some kid with sticky fingers daring to try to touch them.

Within a few days, it was obvious that Mum was struggling with her concentration. She took longer to spot the birds that were clearly visible to everyone else. Dad and Andrés were spending precious minutes pointing out exact locations in the trees, which isn't easy in the jungle. Exchanges such as "Look at the green leaf to the left of the pale branch . . . not that pale branch, the one above it" were common. Mum's frustration began to affect the easy rhythm of our small team as Dad's patience was tested—which is something, given he has the patience of a saint.

Andrés remained serene, however; he was familiar with every species of birdwatcher, from the quietly appreciative Brits, who responded to the rarest of birds with a "What a *nice* bird," to the more expressive Americans, who'd react with a "Wow!" or an "Awesome!"

However, at the same time, there was no stopping Mum; she remained determined to see what she could, despite her growing irritation. The dawn expeditions continued, but it was a night outing that really turned things around for her. Dad had planned a midnight excursion into the forest with Andrés to try to see the Foothill Screech Owl. Mum was going to stay behind with me; Dad had declared that if I spent the whole night looking for an owl almost as mythical as Harry Potter's Hedwig, I'd be fit for nothing the next day. But I had caught the scent of an adventure, and nothing was going to keep me from it.

The Foothill Screech Owl is a species so rare and difficult to see that even Andrés had never glimpsed one. He remained hopeful, however; maybe tonight was the night! And so, at midnight, we found ourselves back in the 4x4, driving away from the lodge and toward dense woodland.

Up to this point, owling, like sea-watching, had not been my favorite way to spot birds. Given that I refused to nap during siesta time, my energy for nighttime outings was generally on the low side, but Dad's resistance— "An eight-year-old needs her sleep"—had made me belligerent. I was a birder, part of a birding family; we were in this together.

A few kilometers beyond the lodge, we parked on the side of the road and walked into woodland. There was no moon and no stars; our chances of spotting a rare bird in a black forest were low. I was beginning to think about my soft pillow when the Foothill Screech Owl's distinctive lilting trill (not a screech!) startled us.

Dad and Andrés immediately swung their torch beams into the trees, scouring the high branches inch by inch. Owls, when confronted with bright lights, tend to behave like deer caught in a car's headlights: they are much more likely to freeze than flee. They are also fantastically camouflaged in the trees, and their eyes gleaming in electric light is what helps you spot them in the first place. Mum and I waited for a long time. The owl did not appear. Its call grew fainter, melting into the background cacophony of frogs and insects, and our initial excitement turned into frustration. We could still hear the owl, but we couldn't see it. Dad continued to scan the trees and I decided to go back to the car, a couple of meters away from where we stood, to wait until they decided to give up and go home. The moon came out, shining into the treetops on the opposite side of the track and illuminating the empty road.

"Mya! Come back," hissed Mum, running up behind me, Dad and Andrés on her heels; the Foothill Screech Owl's shy call had inched closer to our small party, and they didn't want me to miss it. Its song was louder again and moving toward the road. Toward me.

At that moment, the headlamps of a truck blazed onto the road, catching us, immobile, in their full glare as it lumbered toward us. Turning around, I glimpsed Dad and Andrés shielding their eyes from the lights; beyond them, above them, something was flying out of the trees. I slowly raised my hand and pointed at the silent white shape moving across the sky and over the top of the truck. Mum gasped, grabbing my arm. But Dad and Andrés had missed it, hands still shielding their eyes.

Meanwhile, the truck was staggering along, delaying my desperate dad from crossing the road to carry on searching. But the owl had stopped calling, and no amount of torchlight beamed into dark branches and scrubby bushes would reveal it; the Foothill Screech Owl had flown away.

Eventually, as we headed back to the car—Dad and Andrés trudging, miserable; Mum and I trying to hide our grins—the men had to concede defeat. Dad is an amazing birder with an uncanny ability to sniff out a bird in hiding, but for the first time ever, I had seen something extraordinary, *and he hadn't.* I could see that Mum felt just as smug by the way she was grinning; her irritation had lifted.

It was the journey into the Amazon rainforest that remains clearest in my memories of Ecuador. I felt as if I were in an Indiana Jones film as we sailed down the Napo River to the wildlife center where we would be staying. The center was an ecotourism project set up by the local Indigenous population to raise funds to support both their landscape and facilities such as schools and hospitals. As part of a massive conservation effort, they had abandoned hunting on their tribal lands to create a welcoming habitat for the birds and wildlife. Many Indigenous peoples in the Amazon survive by logging, oil mining, or ecotourism: these are their choices. Those who choose ecotourism rely on foreigners visiting the country; if people stop coming, they will have no choice but to turn to logging, which, it is argued, will make a far worse impact on our climate than traveling by plane.

Through binoculars I peered between the cracks in the thick vegetation, branches stretched across the wide river. I had never imagined the rainforest to be so vast or so overflowing with the sounds of wildlife within its leafy walls.

We watched kingfishers and herons leave their posts on the banks of the river to dive into the water and fill up on the small fish placidly swimming about.

The wildlife center was on the still blue waters of Añangu Lake, in the midst of the thick greenery of the forest. A flock of oropendolas was poised on the thatched roof of our lodge, welcoming us between bursts of activity as they wove their hanging basket nests, which dangled precariously from the ends of overhanging branches.

Mum and Dad went off to take a siesta—by now we were in sync with the rhythm of the birds. In the Amazon, they are most active between five and ten in the morning, after which they have stored enough calories to take a nap and shelter from the sun. They're back again for a short burst in the late afternoon, before roosting for the night.

While my parents dozed, I decided to explore, always on the lookout for that elusive Harpy Eagle. Instead, I came upon a group of Hoatzins moving noisily through the riverside trees, grunting and groaning their way through the low canopy. I felt as though I had discovered the lost world of the dinosaurs. Surely this was what an archaeopteryx would have looked like 150 million years ago? Their long necks supported small, blue-skinned, unfeathered faces and scary maroon eyes, topped off by spiky red crests. The young of the species actually retain claws on two of their wing digits to help them scramble among the branches; the Hoatzins aren't great fliers. An unpleasant odor wafted across the clammy, humid air. I discovered later, from our guide, that the birds were known locally as "stinky turkeys" because of the smell coming from their specially enlarged crops, the place where they store food before it is digested. In the crops, bacteria ferments the leaves they've feasted upon, releasing a very shitty smell. With a ruminant's digestive system, they are a little like cows with wings.

We entered the third and final week of our trip, and there was still no sign of the Harpy Eagle. While Dad kept telling me it wasn't too late, it was hard not to feel despondent. We were on our way to Paz de las Aves, decidedly *not* noted for its Harpy Eagles. Tandayapa Bird Lodge—situated in the Andean cloud forest, where misty clouds sweep across the upper canopies of woodland and mountaintops—was altogether much more like Middle-earth than our earth.

Antpittas are dumpy-looking birds, known among birders as "skulky" animals, which are usually brown and obsessed with hunkering down in their equally brown habitats, making them very difficult to see or identify. Ángel Paz owned an antpitta sanctuary very close to where we were staying; obviously, it was famous for its antpittas, not its Harpy Eagles, so I wasn't hopeful for this next stop on our trip.

Before it became a hot spot, Ángel Paz's bird sanctuary was a private farm, which he inherited with his siblings. Paz fell in love with the antpittas that congregated on his land, and he converted his section into a nature reserve and birding sanctuary, while his brothers cleared theirs for farming cattle. His sanctuary became popular with birders the world over, and vastly more successful than cattle farming.

Wandering the ancient woodland, we saw six different species of antpitta, including the rarest, the Giant Antpitta, a bird of legendary status and extremely hard to see. As its name suggests, the Giant Antpitta is a large member of the antpitta family, ranging in length from twenty-four to twenty-eight centimeters.

Ángel Paz spent years caring for these birds and had gone so far as to give them names. The weird thing is that they responded to these names. The Giant Antpitta was María. We watched Ángel call her, and in seconds she came bounding out of the undergrowth and onto the path in front of us. A mostly brown bird with a bright copper underbelly, she dipped her head to feast on the wriggling worms Ángel had provided for her.

But it was the Ochre-breasted Antpitta I was most desperate to see, and she was the last to put in an appearance. Deep in the woods, after we'd spent hours waiting and watching, a bird not unlike a baby robin flew into the lower branches of a tall tree, the upper branches doused in the ever-present shifting mist. Ángel called, "Shakira," very quietly, and the tiny bird, a feathery Ping-Pong ball with a head, cautiously approached his outstretched hand and was rewarded with a worm. She was a fun-size marvel, from her ocher throat to her pale tummy. There was something about the way she moved that was mesmerizing, swinging her body from side to side as she fed. I wanted to fold my fingers around her body to feel her heartbeat, but she wasn't coming anywhere near the rest of us. She only had eyes for Ángel.

Ángel had named this particular Ochre-breasted Antpitta after Shakira, the Colombian singer famous for her song "Hips Don't Lie" and, of course, for her booty-shaking dance moves.

While Shakira was a highlight of the trip, we were defeated by the Harpy Eagle. I sat for hours in the various canopy watchtowers of the lodges we had visited in the hope of seeing one, but none flew my way, and the watchtowers were often precarious places, a mixture of ropes and creaking walkways. On a few occasions I lost my nerve on these sky-high platforms, imagining a rotten board giving way, plunging me into the cloud-covered forests below. It would be another nine years before I was to see my first Harpy Eagle.

Despite the highlight of Ecuador, 2010 was a tough year. The trip *was* good for Mum overall, and in the final days, she had even started talking about going back to work. As we moved between lodges, spotting our target birds, it was easy to forget she had ever been sad or that Dad had felt overwhelmed. The pressure of their lives at home disappeared—for a while, at least.

The benefits of the trip didn't immediately vanish when we returned. Mum had thrived while we were away; her drive to see beautiful birds had cleared her mind to some degree, and this clarity lingered. Dad could see how much better Mum was and made an announcement. From now on, we would change the way we lived. There would be no more unnecessary spending; if traveling was what kept our family together, then everything we had or made would go toward such trips as Ecuador. It wasn't just Mum who improved when we were away; Dad needed the break as much as she did. And they couldn't leave me behind, could they?

Dad was prioritizing experience over possessions; it was to become his mantra. He was choosing mindful birding over suffering. World birding is a busy business, and whether you're in the rainforest or the savanna, there is so much to look out for, to focus on, to wait for, that there is literally no time to think about anything else. In this way, Mum might recover in increments and Dad would regain the mental space he needed to help her.

After Ecuador, we began to live very simply, saving every penny for our travels. I understood, even as a kid, that large and small—brown, patterned, jeweled, or featherless—there was something about birds that made us, even just for moments at a time, lift our eyes away from our lives and up to the skies.

{ 4 }

The Uninvited Guest

YELLOW-HEADED PICATHARTES

The Yellow-headed Picathartes lives in the tropical forests along the coast of West Africa. These birds also go by the name of Bald-headed Crow, for their bizarre, featherless yellow-and-black head, and White-necked Rockfowl, referring to their habit of communally nesting on rock overhangs and cave walls, where they attach their cup-shaped mud nests. The Yellow-headed Picathartes feeds primarily on insects. One of its foraging techniques involves tracking large army ant swarms through the forest, picking off insects that flee the advancing colony.

The birds' forest range in Africa is being destroyed at an unsustainable rate, and population numbers are rapidly dwindling. In Ghana, the Yellow-headed Picathartes was only rediscovered in 2003, after an absence of forty years. Subsequent work with the local community has ensured the colony is protected, with a growing population of birds, and the community itself has benefited from entry and guiding fees and the provision of a new school, courtesy of conservation funds.

While Dad's optimism was exactly what Mum needed to hear, it made little difference to how she felt *right now*. In those cold, early months of 2011, she was worried about the fact that neither of them was in work. A note on the MECH list showed that in November, Mum's citalopram increased to 40 mg. She remained signed off from the law firm. Meanwhile, Dad had to take on the role of carer, the primary caregiver for both Mum and me, doing all the practical tasks they had previously shared between them, while also supporting Mum through a mental health crisis that neither of them understood. At the same time, he was trying to shield me from the worst excesses of what was going on and keep my routine as balanced as possible. I wasn't aware of how serious things had become.

In January, on one of her more energetic mornings, Mum declared a trip to Bangladesh to visit family was what she needed. A change of scene, the loving company of aunts and uncles, and a dose of sunshine would maybe flip the switch in her head, and she would come back to us refreshed and recovered. Dad and I would join her a week later. Dad was keen to stay on after Mum and I returned, as he was involved in a project to save the extremely endangered Spoon-billed Sandpiper, now wintering in Bangladesh.

When Dad and I arrived in Dhaka, hoping to see an improvement in Mum, she had seemed happy enough, but her family told him a different story. Mum had been rude and aggressive toward her siblings, and now they, too, were extremely worried about her. Because she wasn't sleeping, she was almost delusional, proclaiming loudly and regularly that she was the best lawyer in the world. She had been helping a relative with his will as well as organizing Bangla lessons for Dad's arrival; neither of these projects came to anything. Dad turned up to an empty classroom, and the will was never written.

Would she be able to get on a plane at the end of the week? Everyone was counting on Dad to sort her out. And to a certain extent, his presence, *our* presence, had made a difference; at least she was getting some sleep now.

The three of us spent a few days together sightseeing and, of course, birding, and Dad thought that Mum seemed well enough to make the trip home; he would take a few days to himself. "The Spoonbilled Sandpiper needs you," Mum told him.

So Mum and I flew home without him.

And for a day or two, everything was fine. I went to bed at eight o'clock each evening, and when I got up in the morning, Mum was busy around the house. It didn't cross my mind that she hadn't gone to bed at all. But when Ayesha and Laila moved in, "just to keep an eye on Mum," I knew something was up.

One day, I came home from school to find Mum juggling metal amulets stuffed with small coils of paper. They were *tabiz*, special Muslim prayers to protect you from dark magic and possession. She placed one beneath her pillow; another hung around her neck from a thin gold chain.

"They're from your *nanu*," Mum told me. "To help me feel better."

"Don't you know, Mya?" my sister explained. "They're to banish the jinn!" Both Mum and my sister burst out laughing.

"Will they help you sleep?" I asked, confused. This was the extent of my understanding: if Mum didn't sleep, she would be sad, or extremely happy, but never just "normal."

"Maybe," Mum said with a grin.

This was how my *nanu* showed Mum she cared. She didn't really understand mental illness, believing that it was a "defect" that would only bring shame on our family. Better that Mum should be possessed by a demon that could be ousted than suffer the label of "crazy."

Ayesha eventually went home, and then it was just the two of us. I was sent off to school each morning with a couple of quid for school dinners instead of my usual packed lunch. Mum was hyped up, talking fast, flitting from subject to subject. I never once saw her go to bed or get up. When Ayesha returned, I heard her on the phone to Dad, telling him that Mum had been using Facebook into the early hours every day since she had returned from Bangladesh, evidence that she hadn't slept at all.

Given that she had not yet been diagnosed with bipolar disorder, my parents had no language to describe what was happening to her, no idea that traveling across different time zones, coupled with disrupted sleep, could trigger a strange and wild episode of illness.

It wasn't until I entered my teens that I learned that this period was a precursor to the catastrophic breakdown Mum would suffer later that year. Her thoughts were growing darker, more compulsive, despite my sister's diligence; Mum had decided that maybe cannabis might calm her nerves, and then declared that, no, cocaine would do the trick, maybe even heroin. Of course, this all came to nothing. A full week of spiraling thoughts culminated in a new fixation: that Mum should stab herself with a kitchen knife. She was desperate for some sort of relief from the constant noise in her head.

Today, Mum is very clear she was living through a psychotic, or "mixed," episode, where bipolar sufferers experience both mania and depression at the same time. This is an extremely dangerous space; Mum was visualizing, very clearly, the violence she wanted to inflict on her own body.

Meanwhile, Dad had arrived in a remote area of Bangladesh to begin birdwatching and was fully expecting to stay with some of Mum's relatives. He turned up to discover the family had gone away and the house was shut up. There were no hotels, and Dad's patchy Bangla wasn't getting him very far with the locals. He called Mum, who alarmed him with her rambling, and in the end, it was my *nanu* who sorted out a place for him to stay. But Mum's incoherence had unnerved him, and after one night and no birdwatching he made his way back to Dhaka, hoping to catch the first flight home. Mum had booked all our tickets through a Bangla bucket-shop travel agent, and he wasn't optimistic she would be up to changing his flight. This time Ayesha came to the rescue.

On the day he returned, Dad took Mum to her GP, who reduced the dosage of her citalopram, deciding her "episode" was a bad side effect of the drug.

While Mum didn't have a name for what was happening to her, she remembered feeling exactly the same back in 2000. Ten years earlier, she had been prescribed a similar drug, Seroxat, also one of a group known as selective serotonin reuptake inhibitors (or SSRIs), to treat her depression. An antidepressant, in 2000 it had been heralded as a wonder drug. A few months into the course, Mum became suicidal. At the time, this delusional thinking was also considered to be a side effect of the drugs. Slowly, while enduring excruciating withdrawal symptoms, she had weaned herself off Seroxat. Today, it is well known that these drugs can trigger manic or "mixed" episodes in those with bipolar disorder.

Mum was definitely suicidal again, and despite the lower dosage, thoughts of killing herself did not go away.

In spring 2011, the insurers of Mum's work sickness policy asked her to attend a private psychiatric assessment. During the session, Mum explained how invincible she felt at times, how despairing at others. The psychiatrist suggested her mania had been triggered by the SSRIs. But how, Mum asked, when she was no longer taking them?

"Once you've had mania as a side effect of the SSRIs, you can't put the genie back in the bottle," he told her. "You are clearly bipolar."

But he wasn't her doctor, and Mum needed an official diagnosis. He advised her to visit a private psychiatrist. Afterward, when she went to see her GP, she was met with resistance. He didn't believe Mum needed a private consultation and so wasn't prepared to refer her.

By this point, Dad was exhausted again. He knew that Mum thrived while we were all away together. She wasn't well, but whatever her condition was called, he also believed that with something to look forward to, a new focus to absorb her attention, she might feel better.

Was it irresponsible to plan a trip while Mum was so ill? Maybe. Mum's family certainly believed Dad wasn't thinking straight, jinn or no jinn. But he was desperate and booked a family birding holiday to Ghana for the following February.

Soon, I was pulling my bird guides off the shelves to begin my Ghana target list. It didn't matter that the trip was more than six months away. I was making notes and drawing pictures of the strange and rare Yellow-headed Picathartes, our number one target bird for the trip. But there were also barbets and batises, hornbills and turacos, so much to look forward to.

Mum had sunk into a depression, but I have to admit I preferred this mood to "manic Mum." I didn't recognize it as depression; to me she was just calmer, less hyper. I left the house in the mornings, believing she was having a lie-in, only briefly waking to plait my hair and wish me a good day. Because she was up when I got in from school, I didn't know she had been in bed in all day, moving slowly around the house, not saying much. If anything, she just seemed very tired to me. This wasn't to last, and within a couple of weeks, she was in a rage. Her manic moods sometimes manifested themselves in a dynamic character—she was chatty, pleasant to be around. At other times, she was furious with everyone and everything.

She had just received an emergency referral with a new team of doctors: psychiatrists who decided she didn't need to be committed, or sectioned, right now but would benefit from daily visits from a mental health "crisis team." During these sessions, Mum would calmly explain how, why, when, and where she might kill herself. These were presented as totally rational arguments with irrational starting points.

"Ayesha has Laila, she doesn't need me," Mum would explain. "I feel undervalued at work, they don't need me. Chris can look after Mya, isn't he doing that anyway?" Mum felt she was taking more emotional support than she was providing, and it would be better for everyone if she wasn't around. Even when she was told, "If you kill yourself, there's a greater chance one of your kids will kill themselves," it failed to penetrate.

Of course, I was unaware of these meetings and how obsessed Mum was with suicide. Around *me*, she was always a bit "too happy," as though she was making a massive effort just to be "normal" . . . and had overshot.

Beneath the cheer, she was trawling the internet for ways to end her life, while Dad explored cyberspace to try to understand where her obsessional thinking came from and what he could do to save her. It became a game of cat and mouse, Mum determined to end it all and Dad equally committed to stopping her.

And then a bird triggered a series of events that would land Mum in hospital.

A week earlier, the call went out that a White-winged Scoter had been seen and identified on the sea among a large scoter flock near Aberdeen. This was a first record for the UK; even during these chaotic months, the lure of a rare bird was very strong. Mum, distracted for a moment from her spiraling thoughts by this unique opportunity, was also excited and, despite Dad's hesitation, was keen to make the long journey north at the weekend. However, just as we were about to pile into the car, she announced that maybe it *was* too much for her, but that we should go ahead without her.

For reasons that were unclear to me, the trip was abandoned. Dad stormed back into the house, and within minutes I was on my way to a hastily arranged sleepover.

It would be years before I learned what had happened that day. And when I did, it felt like a 3D jigsaw puzzle slowly coming together.

Mum's last-minute change of heart had been part of her plan all along: to get us out of the way so she could kill herself. Dad and I would be in Scotland, and now Mum had not only the means and the motivation but, with us away, the space as well.

Dad had immediately seen through her plan and drove Mum directly to her doctor's office. While she was completely absorbed in her delusional thinking, Mum was also frustrated, unable to see a way out of her despair. She had effectively handed the responsibility for her welfare over to my father, who had to decide what happened next.

A psychiatrist was summoned by the GP. Mum needed to be in hospital, he confirmed, whether she consented or not. Dad reluctantly agreed. He later confessed that handing over her care was the hardest decision of his life, but he knew he couldn't keep her safe anymore.

In Bill Oddie's autobiography, *One Flew into the Cuckoo's Egg*, he discusses coming home one day to find that his mother had been taken away to the psychiatric ward and not understanding what had happened to her. Just like Bill, I came home from school, still smarting a little from our abandoned trip north to see the scoter, to find Ayesha there, but not my parents. When Dad finally came home, he told me Mum was going to hospital and not to worry; she was ill, but she was going to get better.

Ill and *better*—one always followed the other, but so far I was yet to see "better" for any substantial length of time. As I was leaving the house to spend the night with my sister, Mum arrived home, a passenger in a strange car with a woman I didn't know. Too dazed by what was going on to register our presence, she stared unseeing through the windshield as Ayesha drove me away.

I discovered later that the stranger in the car was a psychiatric nurse, tasked with supervising Mum while she picked up some clothes and toiletries before she was taken to hospital and then escorted into a secure ward. The MECH list simply states, *Helena goes into hospital after being sectioned.*

That evening, armed with felt-tips, I sat down to make her a card with a simple "get well" message decorated with bunnies and kingfishers. Dad had presented Mum's hospitalization to me as a good thing; sick people went into hospital so they could get the treatment they needed. And initially, I felt reassured, but when the days turned into weeks and I still wasn't allowed to visit her, I began to feel uneasy. It was a full month before I would see Mum again.

Dad, Ayesha—everyone, in fact—spoke in half truths. No one told me exactly what was wrong with Mum, just that she was "ill" or "sick" or "poorly." That she was receiving "treatment" and that she was "where she needs to be."

It wasn't until I visited Mum myself that I got a sense of what was going on, and I didn't like it. The hospital was vast, with long anonymous corridors that smelled of toilets and disinfectant. Dad and I were taken to a room in which Mum sat alone. The door was locked behind us with a loud click. A nurse was stationed outside for the duration of our visit. What was going to happen in this room that required a locked door *and* a guard? Fluorescent strip lighting bleached the color from Mum's face, but she seemed mostly normal, a bit detached maybe. I didn't understand why she couldn't just put on her coat and go home with us. Of course, she wasn't "normal"—she was just using incredible willpower to hold it together for me. This was the beginning of the burgeoning realization that something invisible was at play: If I couldn't see what was wrong, how would I ever know she was truly better?

Mum was to remain in hospital for seven weeks.

The responsibility of having to section his own wife was immense for Dad. There were many nights he cried himself to sleep, often with huge relief that he no longer had the sole burden of trying to keep Mum alive, minute by minute, hour by hour.

Mum had insisted we carry on birding because it made her happy to think of us on the Mendip Hills, binoculars in hand. But we also went farther afield. It felt strange to go birdwatching without her, but at the same time I looked forward to these trips, as they were golden moments away from my Mum-less house.

We drove to the far west of Pembrokeshire in Wales, to St. Justinian, a small harbor where you can catch the boat to Ramsey Island. Our target was a Lesser Gray Shrike, a new bird for me but one that Mum had seen already, which somehow made it feel less of a betrayal.

The Lesser Gray Shrike is a rather smart black, white, and gray bird with a hint of peach across its breast and a bandit-mask face pattern; its heavy, hooked bill gives it the appearance of a mini bird of prey. The bird showed well, the weather was beautiful, the coastline stunning, and we breathed in fresh, clean air for what felt like the first time in months.

Dad and I strolled around the headland, taking in the common birds: A pair of stonechats, their call mimicking their name, the male black-capped, white-collared, and with a breast of burnt orange, denizens of the heath. Black-and-white Razorbills and Common Murres, which bobbed up and down on the sea, their wings designed for swimming under the water rather than flying. (While they typically seek their fishy food twenty-five to thirty meters below the surface of the sea, they have been recorded at ocean depths as profound as an astonishing 180 meters!)

Auks, as they are collectively known, nest in vast colonies on the sides of sheer cliffs, often on rocky islands. Murres nest at incredible density, up to twenty pairs occupying one square meter of cliff face. Not bothering to make a nest, they simply lay their eggs on narrow ledges. Razorbills are a little fussier and prefer to seek out a convenient crevice. The sight, sound, and overpowering smell of these colonies have to be experienced to be believed.

Enormous Northern Gannets, flashing white in the sun as they plunge-dived into the water like torpedoes, passed by. We smiled, we laughed, we briefly basked in the light, forgetting, for a moment, the shadow over our lives.

July 24 is a very poignant day for my dad; it is the day his own father died, unexpectedly, at the age of fifty-three, when Dad was just twenty-seven years old. A lover of the countryside and a keen, but not obsessive, birdwatcher, my granddad handed down to Dad a deep love of nature. Dad's fondest memories of his father feature them walking the fells of the Lakes or the Yorkshire moors, armed with flasks of coffee and binoculars.

When the anniversary rolls around, Dad is quieter, more contemplative, wondering what his father would think of his life choices. None of us is immortal, and inheriting a family had focused Dad's mind at an early age. It was later in life, however, when Mum became ill, that Dad recalled another of his father's convictions: to always value experience over possessions.

This year the anniversary weighed heavier still with Mum away in hospital. Distraction came in the form of a Stilt Sandpiper at Lodmoor, an RSPB reserve abutting the coast road on the outskirts of Weymouth in Dorset. We watched it in the early-evening sun, as it displayed its breeding plumage. Perched up high on a bank and staring through a telescope, I could see it probing the depths of the marshy pool with its long bill. The rusty ear coverts, patches behind each eye, and a series of dark bars on its underside made it easy to pick out among the more common waders, even at a distance. A species that should have been migrating south through North America at that moment was instead mirroring that instinct in Europe, heading onward to Africa, having been blown here over the Atlantic in a previous autumn storm. When I had had my fill of the sandpiper I turned to face the sea and watched the Sandwich Terns rising and dipping over the breaking waves.

I enjoyed these expeditions—enjoyed adding to my lists, making notes on my "best" birds, and hanging out with Dad—but I was desperate for Mum to come home. Birdwatching without her truly had begun to feel like a betrayal.

Mum was eventually discharged, and for the next few months she was caught up in a series of consultations with various psychiatrists who were still not prepared to provide a diagnosis. This sounds weird, but it isn't uncommon. This all happened during a period of peak austerity; the NHS was in crisis and mental health services had been slashed. They would treat the symptoms only as they presented themselves; no one was prepared to treat Mum for a condition she "might" have.

They believed she was depressed, and that was their focus. Mum believed her time in hospital hadn't changed a single thing about the way she was feeling, merely postponed an inevitable crash. I was relieved she was home, but if she had been ill before, I couldn't see how she was better now. Often in bed when I came home from school, Mum wasn't very interested in either me or Dad.

One evening in September, Mum sneaked out of the house and drove away. A little later, realizing both she and her car were gone, Dad's first instinct was to follow her; she wasn't in any state to be by herself. But how could he? I was asleep upstairs. Dad proceeded to phone around to the family and their friends, growing ever more anxious with each exchange. Eventually he called the police, who began to check CCTV footage and whether she had used any of her bank cards.

Mum, meanwhile, had driven to Chew Valley Lake with the intention of plunging the car and herself into the water. She didn't answer any of Dad's increasingly panicked calls, and then she turned off her phone so she couldn't be tracked. She sat for a long time, trying to gear up the courage to drown herself. But the longer she waited, the more immobile she felt. When a police helicopter began to circle overhead, she drove to a nearby wooded lane, where she hid beneath the trees.

She thought long and hard about what she was doing; she wasn't going to kill herself that night and she certainly didn't want to be sectioned again, but this was exactly the sort of behavior that might land her back in hospital. She arrived back at home to find a couple of squad cars in our drive, and panicked that she might be arrested for wasting police time at the very least. But the police were kind and very understanding.

(I was to learn of this episode two years later. By then the gravity of the event had worn off for Mum and she presented it to me as a lighthearted anecdote, while I listened, shocked and incredulous.)

Later, Mum complained to her NHS psychiatrist that she believed her new SSRI prescription was responsible for the suicide attempt and for her increasingly erratic behavior in general. She wanted different meds. Mum was accused of being noncompliant; she had to take the pills she was prescribed. Meanwhile, Dad, once again, was tearing out his hair, at a loss to know what was best for her.

Older members of Mum's family were still pursuing the idea of demonic possession, rather than mania, as the cause of her illness. There was an aura of shame around mental illness in our Bengali community, who reasoned that the less the family spoke of it, or even acknowledged it was a real condition, the less other people would look at us with fear or, worse, pity. This was their attitude despite the fact that bipolar disorder is frequently inherited and others in our family might suffer the same fate, including me.

That winter, on a particularly dark, wet, and cold evening, my parents sat me down and told me that Mum wasn't getting any better and the best thing for everyone was a long birding break. To my parents, at least, this seemed like a perfectly rational reaction to a difficult situation.

They decided six months in South America was just what we needed. The Big Year had shown us that birding, and being in nature, was good for Mum's mental health, and Dad's too. Our family was falling apart; if we were going to survive, we needed to do something dramatic. And anyway, Mum was at her wits' end, prepared to try anything.

Dad's golden handshake was still in the bank, enough to fund this trip and hopefully future family birding trips. I'm a lucky girl, aren't I? There's no way to measure the impact of a suicidal mother on her children, or of having a father preoccupied with keeping her alive, so if I'm lucky, it's because we could take time out of our lives to live on a different continent for weeks, as we attempted to keep the family together, body and soul.

The previous summer, Dad had booked a winter trip to Ghana; my parents decided it would be a good barometer for the challenges South America would present later that year. I remembered Ecuador, the landscape and the birds. Would I enjoy the same delicious feeling of being somewhere entirely different from the Chew Valley? I hoped so.

In January 2012, with some trepidation, we began our twelve-day Ghanaian tour in Kakum National Park, which houses Ghana's largest remaining tract of rainforest. Perched on a canopy walkway forty meters high, I immersed myself in the birdlife in a new and intoxicating way. The dense tree canopy trapped the humid air beneath, but it did, at least, protect us from the full glare of the sun. Just a few thin shafts of light made it through the canopy.

A confusing array of different species of greenbuls, difficult to tell apart for the uninitiated, were joined by spectacular hornbills: Brown-cheeked, African Pied, and, my favorite, White-crested, with its spiky punk hairstyle. A Yellow-billed Turaco plucked figs from a fruiting tree, swallowing them whole. Its white eyeliner looked in severe danger of running into its bright green plumage, and then its flight feathers suddenly flashed crimson as it moved from one tree to the next.

We had shared the journey to Ghana with family friends, who joined us for part of the trip, and while their company was welcome, it was a bumpy ride. There was very little decent accommodation away from the coastal resorts, so we ended up far from the birding sites, which meant we often left our lodgings well before dawn and didn't return until after eleven at night. We would then eat, grab a few hours of sleep, get back in the car, and do the whole thing again, banging along the next rutted road. Mum, thankfully, was keen to be outside all the time; if our friends ever delayed these dawn excursions, they felt the edge of her tongue, as did Dad when she struggled to spot this or that bird.

Mum was getting very little sleep, usually the last to bed and the first up, which pointed to mania. She was trying to hold it together, even I could see that, when she walked away from any brewing disagreement. Very quickly, the others became aware of her triggers, and if we weren't birdwatching together, they would disappear, giving Mum some space. Our guides weren't faring much better, often on the receiving end of a tongue-lashing if she missed a bird or if they insisted on traveling farther afield than planned to try to see a target species.

But it wasn't like this every day. For the most part, we lost ourselves in the routine of seasoned birdwatchers. For the best birding, particularly in hot countries, you need to be on-site at dawn; the birds just melt away into the shade after the first few hours of light. We are so used to paved roads in the UK, but when you're on potholed roads and dirt tracks with deep ruts, traveling barely above walking or jogging speed, it takes ages to get anywhere. Couple this with the remoteness of many of the best birding locations—the reason the habitat still exists is *because* it is remote—and it's no surprise that serious world birding is considered, by some, to be an ultra-endurance sport.

Kakum National Park was a majestic and relaxing way to ease us into the trip, and the canopy bridge, despite its height and rickety structure, was the perfect viewing platform, so every day I filled up on the sheer quantity of birdlife.

That walkway provided me with views of some of my all-time favorite birds. It was love at first sight for the Black Bee-eater, about twenty centimeters in length, known for its striking cherry-red throat and jet-black back, wings, and head, all of which are offset by a belly speckled with a rich scattering of cyan. Its long, slightly curved bill is designed to pluck bees and wasps from the air before thrashing them against a hard surface to remove the sting.

The African Emerald Cuckoo was another visitor to my tower in the sky. The male has iridescent green feathers that provide the perfect camouflage among the tropical shrubs, offset by a deep golden-yellow belly. The term *beady eye* was invented for this bird, its black pupils fixing me with a steely glare. Like many other members of the cuckoo family, it is a brood parasite. Laying its eggs in other birds' nests, it avoids having to bring up its own chicks.

Our trip had been organized by a company committed to ethical and sustainable tourism, which put environment and community first. Supporting community-run businesses, such agencies create income for the local people, thereby reducing the negative impact on local ecosystems due to hunting and farming activities. The guides are usually hired from surrounding villages, and in our case, a percentage of our travel fees went toward building local schools. The Yellow-headed Picathartes' habitat had been preserved by an initiative to protect endangered and vulnerable wildlife species.

Our friends had left to get some time alone, and they were probably glad to have a break from us: Mum had had a big row with them a couple of mornings earlier. She had got our schedule muddled—she thought we were having a five o'clock start, and they believed we were leaving at half past five. It doesn't sound like a big deal, but Mum was furious: she had lost a precious half hour of birding. As a result, we had endured a day of excessive politeness before they left.

So, it was just the three of us—together with William, our guide, and a local guide from a nearby village—who, in the sweltering heat, made the slow and steady climb toward the forest on the north side of the park. Mum seemed happier once we were on our own again, and so was I. I had had enough of tense silences and strained civility.

An approaching villager, gun slung over his shoulder, was swinging a dead animal by the tail as he walked toward us. Dad and I both instantly realized it was a rat the size of a cat. If my rat-phobic mother spotted it, the day would be ruined. Fortunately, the man, perhaps noting the look of horror on our faces, hid the corpse behind his back. But Mum had seen something. "It's just a squirrel," Dad said, forcing a grin.

Inside the forest, we followed William along a winding path thick with leaves and roots. William had worked for the Ghana Wildlife Division for ten years and is credited with carrying out more ornithological research in the country than anyone else. Throughout this period, he kept in his wallet a tattered, decades-old photo of our target bird for the trip—the Yellow-headed Picathartes, a bird that until 2003 had not been seen in Ghana since the 1960s and was presumed extinct.

William told us his story of wandering through the nearby village of Bonkro, one of the many villages he had visited while pursuing that mythical bird. After asking if anyone had spotted the bird, to his amazement, a local hunter explained he had in fact recently seen more than one and offered to take him to the site, only an hour's walk away. William was able to verify the identification and was elated. Not only was the Yellow-headed Picathartes *not* extinct, but there appeared to be a viable population in this rainforest besides. The holy grail of Ghanaian birding had been rediscovered.

It was midafternoon, and as you would expect at this time of the day, bird activity was slow, although we did pick up an African Broadbill on the way, sitting motionless, half-hidden in a tree beside the trail, waiting for a passing beetle or grasshopper to pounce on. A streaky, olive-toned bird with a black cap, the African Broadbill is usually hard to see in West Africa—a great bonus bird. Our focus, however, was to reach the inselberg, a raised rocky area protruding from the hillside in the middle of the jungle, arriving long before the star of the show would, hopefully, put in an appearance. This was to make sure we were quietly settled in place; the last thing we wanted to do was spook it.

Dad began his usual routine of physically *positioning* us in our nook. He wanted to make sure we had a clear view of the bird when it arrived. He was all business now, dropping down to my level, to Mum's, seeing what we were seeing, adjusting us so that that branch, those leaves, that rock didn't obscure our view. This was a familiar routine: Dad couldn't enjoy himself until he was sure we were all going to have the same experience.

Crouching, semi-hidden, Mum, Dad, and William agreed on a series of hand signals to use should the Yellow-headed Picathartes appear. And then it was just a case of patiently waiting, and then waiting some more.

Half an hour went by, and then an hour; still no sign. I began to draw shapes in the dirt. Mum told me to stop it and to stop scratching my head, because (a) I might pick up an infection, and (b) I might frighten off the bird with my fidgeting.

I had no time to answer back because William was indicating, with a subtle flick of his fingers, that a bird was approaching up a ravine off to the left, still out of view to me, Dad, and Mum. This is usually the tensest moment of any twitch—when the bird is both there and not there. Slowly, it picked its way through the trees, and finally, it strode into sight, perching on a low branch while we stared at it open-mouthed. We could only share our wild excitement through eye contact; the pressure not to scare the bird was huge. Another trotted into the clearing, and soon they were pecking at dry leaves on the forest floor.

When none of us moved, the birds gained confidence and hopped closer and closer, until they were only a few meters away. This prehistoric bird (they have forty-four-million-year-old ancestors), also known as the White-necked Rockfowl, is one of the most unusual birds I have ever seen. Around forty centimeters long, slimmer but not dissimilar to a chicken, the Yellow-headed Picathartes's head is featherless and yellow, with black splotches on either side, like giant headphones, and a stout, dark bill and bulging black eyes. All of this sits atop a long, thin neck. It has a waxy white belly, punctuated by slate-gray wings, but the star feature is its spindly, silvery, bluish-gray legs—which look like they've been borrowed from a chicken. The picathartes appeared strangely smooth and unblemished in a way that didn't quite sit right on a bird, almost as though someone had constructed the different parts of its body separately and slotted them together afterward. This is even captured in the etymology of its name: *pica*, meaning "magpie" in Latin, and *cathartes*, meaning "vulture." It is definitely a puzzling bird.

With its peculiar habits and appearance, patchy and inaccessible distribution, and general rarity, it is considered a crowning glory on any birdwatcher's list, and I was the first foreign child to have seen one.

That night, which was about halfway through the trip, I found a bite on the top of my head. It wasn't dissimilar to all the other mosquito bites I had, apart from the fact that it hurt when I poked it. "Stop poking it, then!" was Dad's advice. Mum inspected the lump and told me it was a small red insect bite and not to touch it, confirming in an I-told-you-so voice that it was my own fault for mucking around in the dirt. A day or two later, the lump had swollen. Now, with a torch, Mum peered at it more closely. There was a hair follicle in the middle of the bite, which she removed in case it was the source of infection. Later still, a hole appeared where the hair follicle had been plucked, and was the size of a pea. Dad doused it with disinfectant. We would in any case be home soon and would deal with it then. I continued to scratch and poke at it, but our next stop took my mind off the suppurating mess on my head entirely.

Toward the end of the Ghana trip, we visited Elmina Castle, with its "Door of No Return," a hole in the fortress wall through which kidnapped Africans were transferred to British ships. It was the last place they would have stood on their home continent before being transported across the Atlantic for a brutal life of forced labor. One of the first things I noticed was the imposing, whitewashed walls, and how they didn't give any clue to their interior of dark and dirty bricked dungeons. The tiny cells with no light or air would have been crammed full of desperate people, their first taste of enslavement. By the eighteenth century, thirty thousand persons were trafficked through this door each year.

At nine years old, having grown up around Bristol, a city that built its wealth off the back of the transatlantic slave trade, I was aware of chattel slavery. But nothing had ever really brought home the reality of individuals being turned into commodities, with no rights or human dignity; it was painfully easy to visualize the masses of people standing terrified in the courtyard of this castle, then involuntarily confined to ships bound for America and the Caribbean. The floor was covered with flowers left by the visiting descendants of those long-dead ancestors. It had a powerful effect on me, standing there amid the hushed weeping of other tourists. Maybe this experience planted the seed in me that would grow into a determination to fight for human rights issues and against all forms of racism.

The last couple of days of the trip alternated between excitement for a new bird and bouts of sobbing because the wound on my head was so painful. A doctor's appointment was booked for the day after we returned home.

It had been an intense few days for Mum: the sleepless nights, the tangles with our friends. But this was par for the course, and for much of the trip, she had been focused, just like the rest of us, on the thrill of seeing new birds in a new landscape.

Before this trip, Dad's insistence that these holidays were more than simple birdwatching trips—that they were curative to a certain degree—hadn't made much impact on me. I still saw the Big Year and Ecuador as mini-adventures during which I got to do my favorite thing. It was in Ghana that I became properly aware of the shift in Mum's personality when she was out in nature, among birds. Maybe Dad had a point.

It was Dad who took me to the GP, and it was Dad who ended up permanently traumatized by our visit.

I guessed there was more to it than a mosquito bite when, after ten minutes of poking around my scalp with tweezers and a surgical light the doctor ushered Dad out of the room. When they returned, Dad's face was a shade of green while the doctor's went red, as he breathlessly, happily, told me what was wrong. He was clearly delighted with my bump, an exotic highlight among the medley of coughs and colds that too regularly came through his door.

"There's a maggot living in your scalp. A really big one," he gasped. "It's not a wound at all but a hole, a *breathing* hole." The doctor explained that I'd probably picked it up as an egg when mucking about in the dirt in Ghana. (Mum's I-told-you-so face would be waiting for me at home, I knew.) He cheerfully went on to tell me how interesting it was, as he could see the maggot moving around beneath the skin. He had never seen anything like it, he exclaimed, brandishing tweezers. "Unless, of course, you want to wait a few more days for it to become fully grown and exit of its own accord?"

I did not.

My GP then disappeared and returned with a young female doctor to show her the writhing maggot under my skin. She wasn't nearly so impressed, her face a similar shade to Dad's.

The doctor's initial tactic of poking the tweezers into the hole and yanking wasn't working. The maggot had anchored itself to my flesh. He then remembered, from his training, that Vaseline might be useful. It would suffocate the maggot, which, struggling for breath beneath the thick layer of gunk, would come up for air. My head was duly slathered, and we waited, the doctor once again hovering with his tweezers.

This plan was semi-successful in that a couple of times the maggot did indeed resurface, but once again the doctor couldn't get a grip. By now Dad was sitting down, his face in his hands. But the third time was the charm, and after much tugging, the doctor managed to pull it out of its hole. I cried through the whole experience, not only because it *hurt*, but because I could also feel it *moving*.

The doctor popped it into a test tube, corked it, and, delighted, ran off to share his booty with his colleagues.

I asked for a peek on his return and was disgusted but also strangely intrigued by the sight, feeling a sick fascination for what had been living inside my head. It wriggled and squirmed, desperate to get out. It was also *huge*, at least five centimeters long and a centimeter wide. How was there enough space for it to move around? Needless to say, the size of this legendary maggot has since grown with every telling of the tale.

The doctor's parting shot was encouraging. Like all maggots, my uninvited guest had kept its home in my head clean and free of infection and disease; the hole would heal perfectly all by itself.

"I can't believe it," Dad repeated over and over again on the drive home. "I just had to watch a maggot being extracted from my daughter's head."

5

My South American Sojourn

GOLDEN-BACKED MOUNTAIN TANAGER

The Golden-backed Mountain Tanager is a bird with a very restricted and remote range. Known from only five sites in north-central Peru at elevations of between 3,000 and 3,700 meters, it requires large islands of elfin forest surrounded by grassland to thrive. Even at this elevation, where the human population is very small, the habitat is under threat from agricultural fires set to provide grazing areas for cattle. The Golden-backed Mountain Tanager population is believed to be declining and may consist of as few as 250 mature individuals.

If I trace my fingers lightly over my scalp, I can still feel the small raised scar where a tumbu fly maggot once lived, where my hair now grows with a distinct kink. I feel a little queasy whenever I think about it, or catch the bump when I'm washing my hair. I'm more disturbed by it now than I was at age nine. But Ghana had ramped up my appetite for seeing rare birds and endemic species; an alien worm was not going to stop me.

Two months after we returned home, we would be leaving again, but this time to South America, for six months, where 3,500 species awaited us. To me, this meant, very simply, six months off, and however Mum and Dad dressed up the idea of homeschooling, I chose to ignore the idea of lessons on *holiday.*

Mum was worried I would miss my friends and find it hard to bond with them again when we returned, but I wasn't: there were only ten of us in my class at primary school, and we were a tight bunch. Parties, playdates, and sleepovers would go on without me—when I got home, I would slot back in. Apart from missing Ayesha and Laila, the only real downside for me was that I wasn't going to be in the school play, *A Midsummer Night's Dream.*

We discovered it was very easy to withdraw me from school. A local authority official paid us a visit and listened patiently while Mum and Dad laid out their plans for my education while we were traveling, but the official didn't seem very interested in their elaborate reading lists and spreadsheets of lessons. There were no further hoops to jump through and there would be no checking up on us, even if we had decided to homeschool me in the UK. My teachers thought six months in South America was a fantastic idea; I was doing well, and they thought I would thrive even without any formal learning.

Mum and Dad were relieved the whole thing had been so easy to organize, but they were also a little shocked to discover you could take your kid out of school and there would be no one to check up on you. They had believed we would have to provide regular reports of my progress to demonstrate I was receiving a certain level of education. They could have kept me out of school until I was eighteen had they wished and not taught me a single sum.

But Dad had different ideas. He arrived home one evening in January after meeting with my head teacher and explained the plan. As we moved around Colombia, Bolivia, and Peru, he and Mum would use study guides *and* weave in a local education as we went, encouraging me to write prose pieces about key landmarks, the birds we saw, and the people we met. I would learn the natural history of the countries we visited and, of course, the geography of South America. (I clearly remember a "lesson" on the boiling point of water at an elevation of over four thousand meters in the Bolivian Andes, while drinking coca leaf tea.) My English lessons would be generously provided by our close family friend Digby, who was joining us for the Colombian leg of our trip. I liked Digby and wasn't too fazed by the idea of some informal chats about literature. He was also the keenest birder I knew.

Back in 2008, a couple of months after Ayesha had dropped her bombshell, Mum and Dad went to Venezuela on a birding holiday for ten days. The trip had been booked before the bombshell, and they figured if they didn't go now, they might not get another chance for years. And so my *nanu* came to stay with me and Ayesha, and off they went.

Their "birding" holiday wasn't quite what they had hoped for; this was to be Mum's first foray into world birding, but the gentle itinerary was organized around long lunches followed by longer siestas and not much birding to speak of; my parents were growing frustrated. And then Digby came along. He point-blank refused "siesta time" and often skipped the group meals, preferring to wander off and see as many birds as he could. Within a day or two, my parents were joining him on these self-guided expeditions. Digby was a hard-core world birder—by that point he'd already seen over five thousand species—using this trip to get back into his obsession now that his kids were older. He was an ex–army officer and now an English teacher, and Mum and Dad warmed to his independent spirit.

By the end of the trip, they were thick as thieves. My parents were keen British twitchers before they met Digby, but they credit him with inspiring them to look farther afield to become world birders.

By now I had a Kindle, which I loaded with all the free classics. "Better pack loads of chocolate too," Mum advised Dad as our departure date drew closer. Like most nine-year-olds, I enjoyed a good bribe to get my head down.

When my parents first raised the idea of South America, I hadn't really understood the scope of it, believing we were to travel the *world* instead of just one continent, and anyway, six months sounded like years at the time. I had gone to my room, armed with an atlas and a huge sheet of paper, and for the rest of that day I proceeded to carefully draw up an itinerary that would include twenty countries and all seven continents. The trip would take two years.

The plan included the places I most wanted to visit, regardless of their size, location, or proximity to other countries. For example, I planned three weeks in Egypt, where we would base ourselves near the wetlands along the Nile to spy herons, flamingos, and the Redknobbed Coot. In China—where we'd spend six weeks—I would see my first Chinese Leaf Warbler and the Spectacled Parrotbill, and so on. I thoughtfully included some European cities so Ayesha could visit with Laila; we clearly wouldn't have time to pop home ourselves on this long adventure. But when Mum and Dad presented me with an alternative (realistic) itinerary, one that would stick to just the single continent and three countries, I had to agree that it looked far less exhausting than my own.

The moment we returned from Ghana in January, Mum began preparing for South America, just weeks away. All her excess energy went into helping Dad organize a seamless adventure as she researched travel arrangements, including flights and car hire, accommodation, and local guides. The downside to all this activity was that it proved she was definitely manic again, but at least she wasn't suicidal.

It didn't look like she was returning to work; her sick leave had been extended again and again. She wasn't well enough to work, however much she wanted to. We were lucky in many ways, and unlucky in others. Both Dad and I would gladly have stayed at home every year for the rest of our lives if it meant Mum would be well again.

Dad was hopeful Mum would thrive on the trip, and I wasn't anxious about her, but my understanding of mental illness was fairly weak at this point and still linked to her listlessness in hospital rather than the mania I was more familiar with. She was desperate to put the whole thing behind her and dedicated the final few days before we set off to arguably the nerdiest part of any world-birding expedition: the preparation of "the spreadsheets."

Mum consulted the official International Ornithological Congress (IOC) catalog to compile a list of birds for Colombia, Bolivia, and Peru. So far, so simple, but the IOC list is dynamic, regularly updated as new species are discovered and others become extinct. The most complicated part of the whole operation is when the IOC decides, for example, that one species of bird is in fact multiple species, and vice versa. This is known as "splitting" and "lumping," respectively, and causes problems for the birder whose only desire before a big trip is a reliable record of what has been seen in a given country so they can decide the targets as they travel.

Our lists for South America were very long, as there are 1,900 species in Colombia alone. It was an unwieldy document at the best of times, and more so if you accounted for the subspecies.

Spreadsheets completed, in a final burst of energy before we set off, Mum announced she would blog the entire trip.

On my last day at school, I was overwhelmed when my primary school class gave me a huge BON VOYAGE card. I felt a rush of love toward my friends and set up my first email account to keep in touch while I was away.

South America is home to one-third of the world's birds, boasting almost 3,500 species, including roughly 2,500 endemics. A continent of tropical rainforests, savanna grasslands, variable microclimates, and the high-altitude Andean habitats, it contains a rich biodiversity. Easily one of the most magnetic locations for world birders—no other continent supports as many bird species. And of course, we would have another chance to look for the Harpy Eagle in the Amazon.

We had hired Avery, a twenty-year-old Canadian, to be our bird guide in Colombia. He worked for Fundación ProAves, a nonprofit organization whose primary aim is to protect threatened bird species and the places where they live across Colombia, through research, conservation action, education, and community outreach. Set up to save the Yellow-eared Parrot from extinction, the group caught the attention of Conservation Allies, an NGO that identifies the most dedicated and efficient local nonprofits with a proven track record of major impact. It is because of such NGOs that enterprises like ProAves continue to thrive. They are provided with technical assistance at no cost and a tax-deductible platform to help raise funds for their conservation work. Thanks to generous financial backing, Conservation Allies charges no overheads or administrative costs on donations to any of their partners. It is unique in offering a true partnership for local conservation organizations in the countries of greatest need.

ProAves has a conservation team of fifty-six full-time staff in Colombia and the largest network of nature reserves in the tropics, protecting some 12 percent of all the world's bird species in their reserves and more than 70 percent of Colombia's most threatened species. They carry out direct conservation actions in twenty-two of the thirty-one departments in Colombia.

The critically endangered Blue-billed Curassow was saved because of the ProAves proactive approach, which resulted in a ban on hunting this turkey-sized bird. Consequently, the population density of the Blue-billed Curassow has significantly increased.

In the first decade of the twenty-first century, there was less of a narrative around one's individual choices regarding climate change, namely the impact of your carbon footprint, and this was also true in the birding community. Today, we are all more aware of the impact of our choices than ever, even if we choose to do nothing about it. Because world birding is closely associated with environmental issues, my parents have long believed that if we're to fly long distances, then we must also balance our carbon footprint on arrival in a new country. By using local guides and staying in ecolodges, they have tried to ensure that the local economy benefits and that we generally make a positive impact in the places we visit. ProAves was our company of choice for Colombia. They owned and maintained their own lodges and nature reserves across the country. Some of our fees went to sustaining the pristine landscapes within which endangered species could thrive. Such landscapes continue to prosper, and rare species to survive, in no small part due to ecotourism. To make a positive impact is crucial if we're to fly *and* benefit the local ecology.

Ecotourism provides an income for local people, enabling them to live well without having to take on work that will ultimately destroy habitats, such as logging, cattle grazing, and mining. With a reliable alternative income, they may turn away from hunting animals and fish for food. In this way, animal numbers increase and tourists have a much better chance of seeing their favored creatures in the wild.

To fly or not to fly? In order for eco-projects to survive, they need tourists. The habitats and animals need tourists to fund the critical conservation work and ensure that there is no need to return to environmentally hostile industries. Today I am an ambassador for Survival

International, which campaigns for the human rights of Indigenous peoples. The poorest countries and most vulnerable people—namely the global south, or developing countries—suffer the most. These countries depend on agriculture and will be most severely affected by climate change. This unfairness has led to a new movement: the movement for global climate justice.

If conservation projects are shelved because of concerns about carbon emissions, then people, habitats, animals, and birds will suffer. If Indigenous peoples are shoved off their ancestral lands in the name of conservation, then they will suffer. Some 71 percent of the world's carbon output comes from one hundred companies, and not from the individual flying habits of travelers. As with most things, there is a balance to be achieved during the necessary period of transition. The impact of deforestation causes more than three times the current total emissions of all aviation, as well as untold devastation to Indigenous peoples and the environment, a figure that would undoubtably increase if all ecotourism ceased tomorrow.

It wasn't until we landed in Bogotá that the reality of our trip finally began to sink in: six months of birding. Not even the downpour in which we touched down dulled my fever to get going. We met our guide, Avery, and were reunited with Digby, bearded, slightly rotund, and wearing the ancient fishing vest I envied. It was dark green and covered in pockets; all he had to do was reach inside one random pouch or another to produce an umbrella, compass, map, pocketknife, or whatever else he needed. A good portion of this trip would be spent trying to find an identical gilet to fit me. (We never did.)

It's no wonder Mum and Digby are such good friends. They are very similar in lots of ways: both quite obsessive characters, always desperate to see the next bird (although who isn't!), a little competitive, and keen to pack as much birdwatching into the day as possible. Just don't get them started on politics...

But the trip had just begun; no one wanted to quarrel, and anyway, Digby would be too busy homeschooling me in English to get too competitive with Mum.

The first couple of days were spent birding around Bogotá, including at Chingaza National Nature Park in the eastern Andes, considered one of the great natural treasures of Colombia, with its vast range of mountains shrouded in white clouds, glacial lakes, and jungles of gnarly trees.

We settled into a rhythm, rising early to reach our location before dawn. I added four more hummingbirds to my list, the most notable of which was the very rare and near-endemic Coppery-bellied Puffleg. We were on the mountainside, walking through patches of forests dispersed across farmland and climbing over barbed-wire fences to reach the remaining areas of habitat. We were looking for something else when the Coppery-bellied Puffleg Hummingbird appeared. It was iridescent green with a golden belly, but most intriguing to me were the tufts of feathery cotton balls above its legs—hence *puffleg*.

In Colombia each and every bird was a prize, whether it was the Bogotá Rail or the endemic Brown-breasted Parakeet or the Pale-bellied Tapaculo, the latter causing a huge row between Mum and Digby. Digby had seen it first and, wrapped up in his own excitement at the sight of this tiny, tawny bird scurrying across the ground like a mouse, had delayed calling us for a second. I managed a glimpse when Digby gave a shout, and so did Dad, but Mum had missed it and ranted at Digby for keeping the bird to himself. Digby, understandably affronted, hit back with a few choice words of his own.

"Welcome to the world of birding with the Craigs," announced Dad. This episode was a baptism of sorts for Digby; thankfully it didn't put him off going away with us again.

Mum never did manage to catch the Pale-bellied Tapaculo, and every other day she would bring it up, blaming Dad because he had seen it and she hadn't, blaming both Dad and Digby for bad directions. Her anger was just as intense with each outburst; she's still angry about it today. That's part of the game, though, not just for Mum but for every birder—you miss a bird, you don't stop feeling it.

If there was a downside to these days of amazing birdwatching, it was the car journeys. If I wasn't sleeping, I was being tested from one or another of the textbooks Dad had so carefully packed. He took charge of science and maths, while Digby kept his promise and became my patient English teacher. But it was Mum's lessons that were the most successful, or at least the most pleasurable. While she wasn't quite up to the job in Colombia, by the time we were in Peru, she and I wandered the rainforests and savannas of South America having conversations that never once felt like lessons. These meandering exchanges took in the politics of the region we were visiting, as well as the key historical figures, the food, the music, customs, and wildlife.

My random observations would become history lessons. For example, I spent the entire trip listening to the 2004 album *American Idiot* by Green Day on a loop, inspiring Mum to launch into the stories behind the songs—in this instance, the angst of a generation living in the shadow of the Bush administration, 9/11, and the Iraq War.

While we walked and talked in the shade of forty-meter-tall trees in the moist jungle heat, my "lessons" were interspersed with the stories of my own family history, Dad's childhood in Merseyside, Mum's thousands of cousins, funny stories and sad ones, including the deaths of both their fathers. In the strangest of settings, I felt closer to Mum than I ever had before.

With dawn still an hour away, we left Bogotá in a 4x4 and were soon winding up a steep mountain track. Flooding from the rainy season made the road too hazardous to go very fast. It was still dark, so we were taking it slow and steady when the driver slammed on the brakes, stopping just centimeters away from a massive landslide. To one side of us was the cliff face and to the other a sheer drop into the valley below. The track had been almost entirely swept away. I stared at each of the adults in turn, wondering which of them would decide our next move. From the glint in Mum's eye, I realized that no mere landslide was going to keep her from her birds.

Avery wanted to turn back, but Mum wasn't having any of it. Fuertes's Parrot, our target bird for the day, is a major event for world birders. Endangered in Colombia, it was believed to be extinct until it was rediscovered in 2002, after an absence of ninety years. Mum was desperate to see one, and a risk to life and limb wasn't going to stop her. She urged the driver to maneuver the jeep through the gap in the road. If we were to have any chance of seeing this parrot, we would need to make it up the mountain. The Fuertes's Parrot usually waits for the sun to rise before it swoops out of the forest, into the sky, and away; if you miss sunrise, you miss your bird. We had just a few kilometers to go, but without a vehicle, we weren't going to make it there for dawn.

Eventually, and after wasting precious minutes through in-car fighting, we decided to walk, joining one of many hiking tracks to the top. Mum's determination was infectious, and I ran ahead, way ahead. When I turned around, it was clear that enthusiasm alone wasn't going to get the adults to the top: the stress of the dawn deadline combined with a sudden onset of altitude sickness was taking its toll on them. They were sweating hard and panting.

"Come on!" I yelled. "The sun will be up soon."

"Leave me behind!" Mum shouted dramatically. "I don't want you to miss it. I'll wait here for you." She gave Dad a little push. But Digby took her arm.

"No one gets left behind," he insisted, rallying the troops: once a soldier, always a soldier.

But the sun was rising, light flooding the landslide to reveal farms in the forest valley below and woodland above. I was back in Middle-earth, the city lights of Bogotá behind us. But we were nowhere near the top, and by now, the Fuertes's Parrot would surely have done its dawn flyby.

Giving up, all of us collapsed onto the grassy slopes, silently agreeing we would wait awhile to see which other birds showed up. The air was fresh and sharp, the sky now a mix of orange clouds and blue sky. Our hill was barren, a victim of deforestation, and the few birds we saw were too far away to identify. The morning wore on and we wandered around, aiming our binoculars into the forest below. Would we see even a single bird? None seemed to have shown up for work, and given the huge numbers that live in Colombia, it was a depressing thought.

I was adjusting my binoculars when I noticed Mum standing stock-still, one hand raised in the air, as if she were waiting to answer a question in class.

She was looking through her binoculars into the trees surrounding the undulating fields of farmland. A thick cloud hung over the tallest branches; if Mum was staring at a bird, there would be no way to identify it in all that fog.

"Look!" she said finally, turning around to wave at Dad and Digby. "Over here! Quickly." But they ignored her, so intent were they on their inch-by-inch inspection of the treetops along the mountain ridge. When we were in Ghana, Mum had more than a few times misidentified a species after raising our hopes for a rare or endemic bird, and so Dad wasn't in any hurry to go chasing after birds that weren't there; after all, our parrot had already flown its nest, so to speak. There's a word in our community for birders who claim to have seen a bird they haven't: they're referred to as "stringers," and after a while, just like the boy who cried wolf, no one pays them much attention. Mum wasn't a stringer by any means. She was just as frustrated by her lapses in concentration and her inability to see a bird that was supposedly, according to Dad, right in front of her eyes. But she was sailing very close to the wind of a stringer.

Avery came to her rescue. He positioned his telescope and followed her finger into the trees.

"Fuertes's Parrot!" he yelled, punching the air.

In seconds, Dad and Digby were running back up the hill. Avery gave his telescope to me, and there it was, flying wide circles above the treetops, a flash of color against the green foliage. Critically endangered as a result of deforestation and loss of natural habitat, this rare mossy-green bird spread its red-and-indigo wings, as if to display its coat of many colors just for me, and soared. For a second, the blue of its wings matched the color of the sky, and it flew, like a bullet made of feathers, back into the misty treetops. We cheered!

The fact that we had seen our target bird was a great omen for the rest of the trip. We had earned the doughnuts we enjoyed on the drive back through the traffic jams of Bogotá.

"No more stringer jokes, okay?" Mum warned Dad, her mouth full of doughnut.

He looked sheepish. "I promise," he said. "Do you want the other half of my doughnut?"

The ProAves Cerulean Warbler Bird Reserve is situated on the western slope of the eastern Andes near the town of San Vicente de Chucurí in a vast tract of Colombian oak dry forest. It sits at the end of the historic, mossy, cobbled path of Lengerke, built of stone in 1840 by the German engineer and landowner Geo von Lengerke. Pausing now and then to rest during the steep climb, we couldn't escape thoughts about the passage of time. Thousands of footsteps, maybe millions, had worn these stones smooth. As the seasons changed and the years passed, the rainforest had been plundered for its wood. After it was cut down, it grew back, which all takes time, and during this period of regrowth, efforts to conserve the local habitat also grew and were successful, and thankfully, the birds are back.

We arrived at the reserve very late at night. Miles away from any city or town, there was no light pollution, and we went to bed in the pitch-black, with no idea of our surroundings. We rose early the next morning still in darkness, after just a few hours' sleep, and at three o'clock began a long walk through farmland and up the cobbled path of a very steep hill, trying not to slip on the mossy steps. Dense forest lay on either side of the path; it was a silent, slow, dark trek up the mountain. As dawn broke, so did the birds' nighttime fasting. We watched Gorgeted Wood Quail—rare specimens and usually very hard to see— wobble out from behind the tall oaks to feed. Chestnut-breasted and whistling, three tiny members of this imperiled species hopped onto the track and began to munch on the seeds laid out for them during the night by the site managers.

I was holding my breath for the hummingbirds; we heard the whir of their wings long before we saw them. In a small clearing among towering oaks, low branches had been hung with feeders, and even in the half-light, their psychedelic bodies shimmered. Violetear, Chestnut-bellied, Steely-vented—hummingbirds in their hundreds were lost in a feeding delirium. Into this swarm flew the Black Inca and Indigo-capped Hummingbirds, both new to me and found only in Colombia. I remembered the promise I'd made in Ecuador to see every species in the world, and I remembered why I had made that pledge: very simply, these birds made me happy.

Hummingbirds are amazing creatures. They are found throughout the Americas, from Alaska in the north to Tierra del Fuego in the south. They include the smallest bird in the world, the Bee Hummingbird, a native of Cuba and only five centimeters long. Their wings beat so

fast—up to ninety wingbeats per second in some of the smaller species—
they literally create a humming sound that we can hear. In dense jungle,
you will often hear the birds before you see them or sometimes, frustrat-
ingly, don't see them. It can be alarming if one buzzes past your ear, like a
giant insect. All this wing-flapping is hard work, and hummingbirds have
an incredibly high metabolic rate. Fortunately they have their very own
supply of rocket fuel to power them, the sugar-rich nectar from flowers,
drinking it up through tubes attached to their long tongues. Their wings
are specially adapted to allow them to hover while they sip their favorite
tipple, and, uniquely among birds, they provide lift on the upstroke of the
wing as well as the downstroke. At night, when they cannot feed, hummingbirds can go into a kind of short-term hibernation, a momentary
torpor putting their metabolism on pause to conserve energy.

At midday, we emerged into bright sunlight at the top of the mountain, sweaty and breathless. As we settled down to eat some lunch, our
target bird for the day appeared, and we hadn't even had to look for
it. The endemic Colombian Mountain Grackle soared into view. Jet-
black with an intriguing red patch under its wings, it flew a wide circle
overhead.

Back down the mountainside, in a deep tangle among the shade-
grown coffee the reserve was famous for, we thought we could hear
a Niceforo's Wren. But the tangle was so thick that we couldn't see
much of anything. We followed its voice, our eyes finally alighting on
this wonky-looking bird on a distant branch singing its heart out. Its
enormous bill made it look off-balance and a bit like a predator, which
is funny because it's so small. I high-fived everyone in the group. It was
my two thousandth species and a great landmark bird.

Antpittas are major-league birds in Colombia. They are unique to South
and Central America, with fifty-five species in total; Colombia alone
lays claim to twenty-seven.

They are known for their evasive nature, often proving difficult to distinguish in their natural forest habitats. So it was with no small amount of trepidation and lots of excitement that we visited the remote Río Blanco Nature Reserve, where we had booked a lodge. There was no electricity, and it had rained heavily the night of our arrival. The next day, the woodland was a hazard of slick muddy paths; we couldn't go in until it had dried out, so instead we birded along an open track all morning, and in the afternoon we decided to try again.

Antpittas tend to show up only when they're called. I had already seen the Giant and Ochre-breasted Antpittas in Ecuador, and now, a couple of hours before dusk, we were still hoping to add the Hooded Antpitta to our list. Avery was armed with various antpitta calls on his phone, but not the rare Hooded Antpitta.

The birdsong recording will alert the bird to an interloper on their territory, which, naturally, they will want to defend by, hopefully, flying toward the sound and issuing an immediate evacuation order. Avery reassured us that another guide, out with his group in the same forest, would "give" him the call; on we marched to find him.

When we tracked down this guide in the midst of his own group, he was reluctant to share. Part of the "value" of local guides is their privileged knowledge of the habitat and habits of the birdlife, and in the case of this guide, it was his exclusive ownership of the rare Hooded Antpitta's call. Bird guiding is a competitive business, and it's easy to understand why he wanted to hang on to his unique access code for as long as possible.

It was disappointing, but we birders are generally a stoic bunch; there are always more birds to see! In a sunny spot among towering gumbo limbo trees, we listened to the songs of the tanagers and the toucans, praying they would show themselves.

"Shhh!" Avery was standing very still, pressing a finger to his lips. "Hooded," he whispered.

"Seriously?" whispered Dad, a little desperately.

Avery nodded, and taking out his phone, he recorded the call. Waiting for any antpitta requires masses of patience, so for the next hour we waited, while, at intervals, Avery pinged the call into the forest. By now it was getting dark. Dad was worried about getting lost on our way back to the lodge, but Mum and Digby and I wanted to stay. It was only when Avery, too, decided it was time to leave that we reluctantly headed out.

Emerging, unlucky, from the trees onto the cobbled path, we met up with the other group of birdwatchers again, and Mum began chatting to a couple of the women, describing our hunt for the Hooded Antpitta. They, in turn, told her of their own disappointment: the other guide had played the call over and over, and heard a response each time, but the prize failed to appear. It didn't take very long to work out that there was no Hooded Antpitta, just the bleeps from one iPhone responding to the bleeps from another.

By now, I was growing desperate to see a new hummingbird, so we set off for the Dusky Starfrontlet Bird Reserve, which involved a three-hour journey from our lodge on horseback. Well, I was on horseback. Mum, Dad, and Digby rode mules, their legs dangling comically over the beasts' flanks. This was real Wild West riding: no hats or boots included. Hanging on to the saddles' steel pommels, we climbed steep tracks, rode across grassy hillsides, and waded through fast-flowing streams. The fact that none of us knew how to ride was not a deterrent; on we went, the hummingbird our carrot. Saddlesore, we arrived at the reserve itself, high in the mountain forest, populated by a massive array of hummingbird feeders. The birds were, as usual, buzzing around the food like foil-wrapped sweets. How would I find my prize among so many others? Hot and exhausted, we trekked up to the higher slopes, hoping to spot my bird in the trees. All hummingbirds are gorgeous, but the Dusky Starfrontlet—also known as the Glittering Starfrontlet—has something very special going on. Critically endangered, this tiny bird is notable for the way its rich iridescent plumage of blues, greens, and gold catches the sun and *glitters*.

We waited. My birding lessons from Dad became more focused: how to be very still in a noisy forest, how to position myself to get the best view, where to stand so that I wasn't conspicuous to a bird. Birds are wild creatures, easily spooked; I was learning how to become invisible.

Was that it?

Yes, it was! Perched in direct sunlight on a low branch, twitching and shining, this gleaming bird was too stunning to be real. We watched in silence. The vivid greens of the foliage around us, the piercing blue of the sky, the low-hanging clouds, were all bold, rich colors, impossible not to admire yet barely holding a candle to this dazzling handful of bird, which, characteristically, seemed oblivious to our presence. The more I looked at it, the more I wanted never to turn away.

"You'll be glad to see the back of us," Mum told Digby as he packed: his month in Colombia had come to an end. While Mum sounded as though she was joking, she knew she had been a difficult traveling companion at times. She had occasionally clashed with him, the usual story: she couldn't see the bird everyone else was admiring and lost her temper with whoever was closest to her. Sometimes that person was Digby. But he was also the peacekeeper, often taking the heat out of arguments between my parents. In his final few days, he had come up with the Craig Family Harmony Index, a canny tool for taking the temperature of my family's mood. Generally, we got on well, happy when we had all seen our target bird, except on those occasions when one of us failed to spot it (usually Mum). The index would then plummet, and instead of harmony there would be discord. When Digby waved us goodbye, I was sad to see him and his gilet go. Would his role of peacemaker now fall to me? The Craig Family Harmony Index was fairly stable when he left. We were ready to be three again; we had found our feet.

Our six weeks in Colombia were over. I would miss many things about the country, but not the tough treks up and down the hills. I had to admit, however, that without these long and exhausting days, I wouldn't have seen the endemic species that made this trip a personal birdwatching triumph. I realized that the harder you work for your bird, the greater your reward and the sweeter the pleasure. And...I had added nearly four hundred new birds to my world list.

We flew to Bolivia on my tenth birthday. From El Alto International Airport in La Paz, we headed for the Bolivian Amazon and Sadiri Lodge. Often overshadowed in birding terms by its neighbors Peru and Brazil, Bolivia has on offer an astonishing variety and scale of habitats. The Amazon and the high Andean Plateau, or Altiplano, encompassing the vast Lake Titicaca; the Chiquitano dry forests and open Chaco heading toward Paraguay; the Yungas cloud forests; the rare *Polylepis* forest; and the tall grass savanna of the Beni plains: each environment has its own suite of species uniquely evolved to survive there.

Located inside the vast wilderness of Madidi National Park, Sadiri Lodge is a foothill sanctuary with sweeping views of the Serranías Chiquitanas, and it is ideally placed: the moist tropical forest air acts like a magnet to the birds, and 430 species have been recorded in this area alone.

At the time, I wished we could have stayed at the Chalalán lodge close by, mainly because of the legendary story behind its construction. Israeli adventurer Yossi Ghinsberg yearned to meet writer Henri Charrière, whose memoir *Papillon* recounts his escape from a notorious penal colony in French Guiana to live out his days in Venezuela. Sadly, by the time Ghinsberg had saved enough money for the trip in the 1980s, Charrière had died, but Ghinsberg went to South America anyway.

When a mishap separated him from the three other men he was traveling with in the Bolivian Amazon jungle, Ghinsberg was lost. After three weeks, starving and barely conscious, he was rescued by the Indigenous community of San José de Uchupiamonas. Two of his friends were never seen again.

Ten years later he returned to Bolivia, distraught to find that the Uchupiamona community was shrinking; the young were leaving their families to make new lives in the cities. With Ghinsberg's help, the local Indigenous community built and established Chalalán Ecolodge, which has since become a model for community-led tourism.

It is exactly this type of initiative that was to inspire me time and time again in Bolivia. So many communities had been destroyed because of lucrative legal and illegal logging contracts, their habitats becoming barren and useless. But we also met with those who, just like the Uchupiamona, had taken their future into their own hands and created sanctuaries for the land to flourish and wildlife to thrive.

Ruth Alipaz, from the same tribe, was eleven years old when she bucked the traditions of her community and decided to continue her education. Years later she returned to her people with a dream. She persuaded them to join her in setting up an ecolodge, which was to become Sadiri Lodge, instead of accepting a deforestation contract.

She brought in local tradesmen and craftsmen to train the Indigenous community members to build the lodge and make all the furniture. These skills would continue to earn them a living after the lodge was completed.

Our new guide, Sandro, also from the same tribe, had meanwhile repurposed his hunting skills to become a keen birder. Why waste a sharp eye and a good ear? Ruth's plan worked, and now the lodge provides a sustainable source of income rather than a one-off logging payment and the loss of irreplaceable habitat. In this way, conscientious tourism protects communities *and* endangered species.

Newly opened, Sadiri felt fresh and the area pristine. This was a real adventure; we were Amazonian explorers heading into the unknown. Now I would surely see my Harpy Eagle. Our first trip into the rainforest was marked by a wonder of hummingbirds, resplendent creatures with resplendent names such as Amethyst Woodstar, Blue-tailed Emerald, Golden-tailed Sapphire, all of them jewels, their vivid colors inspiring the scientists who named them. My favorite, the Rufous-crested Coquette, was, at six and a half centimeters, small to say the least, contrasting with the huge Blue-and-yellow Macaws flying overhead. The coquette's black-tipped, spiky orange headdress, another punk hairstyle, made it easy to spot. A Pavonine Cuckoo sat in a tree beside the trail in full view. This normally shy and retiring bird returned our stares. We looked at it, it looked at us, unconcerned; were we the first birdwatchers it had ever encountered? A two-note whistle apparently coming from the undulating forest floor had Sandro stumped: What was it? As we tracked the song through the trees, the call kept moving ahead of us. After a long time, a Ringed Antpipit crept out from among the leaf litter. A ground-dwelling bird and much the same size as a thrush, it pecked at the ground, inhaling insects.

And the Harpy Eagle? We looked, of course. But Sandro had no information on active nests, where the eagle is usually to be found, so all we could do was hope one flew by. We still had weeks ahead of us when we might see the fearsome bird we were all, by now, desperate to catch a glimpse of.

On a tiny plane we flew to the Barba Azul lodge in the Beni savanna. After an hour of cramped flying, we had just begun our descent when

I, the only Spanish speaker of our trio, heard a few choice words from the cockpit. As I was glancing out my window, a few choice words of my own sprang to mind. Our "landing strip" was a field full of cows.

The pilot swooped back and forth over the field, like an oversized swallow, in an attempt to herd the cows to one side, but they barely noticed our tiny plane. After several abortive attempts to land and with our fuel running low, our savior galloped into the field on horseback. The farmer had arrived! He needed only a minute or two to clear the land. Our very relieved pilot let out a few happier curses as he finally touched down.

In a dry landscape—almost the inverse of majestic Sadiri Lodge—and under a pounding sun, we made our first trip to the palm-tree plantations, home of the Blue-throated Macaw, which mostly prefers to nest in the cavities of dead trees. The Blue-throated Macaw is a beautiful parrot, so extraordinary that it became one of the most trafficked birds in the South American pet trade. This, along with deforestation, caused its numbers to plummet. By 1998 there were only thirty-six Blue-throated Macaws in the wild and extinction looked likely, but intensive conservation work has seen its numbers increase to over four hundred birds. It wasn't very difficult to imagine why someone might want one of these birds as a pet: they are stunning. Turquoise blue with a bright yellow belly and a wingspan of almost one meter, these birds look like a slice of sunshine when they take to the air, but to me the idea of caging such beauty is abhorrent. They didn't need to dance or sing or be taught how to talk; it was enough for me that they were alive and wild.

Unfortunately, one of my most distinct memories of Barba Azul wasn't the birds but the mosquitoes. They were thirsty for blood, naturally, but especially keen on mine, and, like homing pigeons, they darted toward me whenever I was outside. And these guys could bite through three layers of clothing. If we weren't birdwatching, I was in our room, spraying myself with repellent before I hid behind the mosquito net.

Between lodges, Mum was blogging, as she had promised. Whenever we had a signal, she transcribed the notes she had made to the internet. Her ambition to chart every leg of our trip had become all-consuming, and Dad was getting more and more irritated by the frequent trips we had to make into the local towns to find an internet café so she could add to her epic online journal. But as obsessions went, this wasn't a bad one, he reasoned; at least Mum was enjoying herself.

Located on Bolivia's western border with Chile, five thousand meters above sea level, is Bolivia's oldest national park: Sajama, home to ice-tipped volcanoes, alien rock formations, hot springs, and old ruins. It felt ancient, reminding me of the early scenes from *2001: A Space Odyssey*, with its images of the earth before man's evolution. The cold, high-altitude temperatures of the Andes were evident in the ruddy cheeks of the Indigenous Aymara population.

Very quickly we understood that breathing was going to be a problem. It felt as if I were trying to walk through a swimming pool with weights tied to my wrists and ankles. How was this going to work? Chatting to a sprightly Aymara farmer, Dad managed to convey the fact that we were having trouble staying upright. Did he have any advice? The farmer pulled a brown paper bag from a pocket and snapped it open to reveal a jumble of green leaves. For good measure, he opened his mouth wide to show us a couple he was busy chewing.

Outside of South America, coca leaves are illegal to grow—anything to do with cocaine is internationally prohibited—but in South America, they have a few traditional, nonrecreational uses, one of which is dealing with altitude sickness. When the farmer offered the bag around, we each took a leaf and dutifully shoved it in our mouths. Later, we were introduced to foul-tasting coca tea; I preferred the leaves, and as long as we chewed them regularly, we were able to keep moving.

But it wasn't just the altitude sickness we struggled with in Sajama—it was also the cold. The lake beside our lodge was covered in a layer of ice every morning, and at night, I had to sleep under a suffocating number of blankets. Few birds can survive these temperatures, and it seemed incongruous that the Chilean Flamingo was one of them. Their long, thin legs looked brittle, as though they would snap off in the frozen waters. Their bright pink plumage clashed spectacularly with the surrounding monotone landscape. Other birds specializing at eking out a living in these extreme conditions by grubbing around on the ground for seeds included the Ornate Tinamou, Least Seedsnipe, Golden-spotted Ground Dove, Taczanowski's Ground-tyrant, and Bright-rumped Yellow Finch. All are adapted to life at high altitude. A mob of Puna Rheas, the South American equivalent of the emu, regularly ambled by on long, sturdy legs. They are flightless birds but able to outrun potential predators, reaching speeds of up to sixty kilometers per hour. Their long necks, large eyes, and sensitive ears act as an early-warning system.

There is invariably a bird you miss, "the one that got away." In this case, it was particularly painful. We could hear the call of the Diademed Sandpiper-plover—a beautiful wader, high on our most-wanted list—floating across the stillness of an Andean lake, but we could not pick it out. I almost had to restrain Dad from stripping off and swimming across to the far shore. "It's too cold!" I pleaded. "You'll die." Reluctantly, he agreed. We still had birds to see. It was far better that all three of us had missed the bird than just one of us.

We left Sajama on the day of a rare astronomical event. On June 5, Venus was to make its transit across the sun. While I was to enjoy only the faintest smudge of a black speck on the sun, I discovered later, and somewhat resentfully, that my friends back in the Chew Valley had seen and celebrated the whole thing. I thought about the sleepovers, playdates, and birthdays I was missing and felt, for the first time, a little melancholy.

There are upward of 1,800 species of bird in Peru, almost as many as in Colombia. With more than one hundred endemic birds in the country, it felt like every new valley we entered had its own unique collection. Even before our plane had landed, we started ticking off birds on the runway. It was literally a birders' paradise. To give some perspective, to date I have seen over five thousand birds—half the world's species—and on this trip to Peru alone, I recorded over a thousand.

By this point we had been in South America for almost three months, nearly halfway through our trip. Peru, with so many more accessible target species to go for than either Colombia or Bolivia, was our last and longest stop. Constantly on the move now, we never stayed anywhere longer than four days. Our routine was chaotic; we were living entirely out of our suitcases, as there was little point in shoving our jumpers into drawers in the morning only to have to pack them that evening. I forgot about my friends, my home, and school; this was my life: traveling, birdwatching, and lessons on the move. Mum had run out of steam for her blog by now, so we rarely had to spend hours tracking down an internet café out in the wilds. It was a good sign; she was relaxing into the adventure, staying in the present. I was aware of the contrast with Ghana, where Mum's frustration with her concentration was a big deal. During our early days in Peru, Mum seemed rested, calmer—despite our constant shifting from place to place—and much happier to take her time spotting birds, even when Dad or I had seen them first.

However, this happy momentum wasn't to last.

It was after Ayesha called us to announce she was pregnant again that Mum's mood shifted. To be fair, neither Mum nor Dad was ecstatic about the news. Despite their support, Ayesha and her partner were still struggling. But I wasn't thinking about any of that—I was just delighted to be expecting a new niece or nephew.

"She's made an active choice to have another baby," Dad said finally. "She's not a child anymore; we have to respect her decision." He was resigned to it.

Much later, I asked Ayesha if she chose to drop her new bombshell while we were away because she was angry about feeling abandoned. "Maybe. A bit," she confessed. But she also understood why we had to go away. She knew, deep down, these trips were about our survival as much as the birds.

In central Peru, we traversed the Andes from west to east as we began our journey to the Bosque Unchog cloud forest. Winding our way along the long, dusty Satipo Road—renowned for being one of the best birding areas in Peru—we were hoping to see as many endemic species as possible, exploring deep valleys and dense forest on our route.

The Peruvian mountainside was flush with birdlife, in pristine habitats on the steep, inaccessible slopes. As we first climbed then descended, a different slice of birdlife was visible at every level.

There was one bird that Mum and Dad insisted on spotting in what felt like every valley and on every slope of our trip. It was the Rufous Antpitta. A couple of years earlier, there were rumors that the twenty or more subspecies of this bird were about to be split into several new species by the scientific community. (Our spreadsheets definitely needed updating.) No one quite knew what would happen, so Dad reasoned it was best to just try to see them all. Everywhere we visited, from the moment we landed in Peru, we looked for the Rufous Antpitta. We even stopped to look for one on a high mountain pass in the middle of nowhere in a tiny fragment of suitable habitat. This one was so obscure the subspecies hadn't even been given a scientific name, but Dad wasn't taking any chances.

The reason for this obsessive behavior is a little nerdy to non-birders. If a species is split, a keen birdwatcher would want to add each of the "new" varieties to their list. We needed to see as many varieties of ant-pitta as possible *now* so we wouldn't have to return at some point in the future to capture the newly reclassified species we may have missed. The key distinguishing feature of each subspecies is its call, which is true for all birds, but unlike other animals, the lines dividing them are thin. The question of whether small differences equal a mild variance within a species or define a whole new species is a hard one to answer. And some are even differentiated solely on their DNA because there are no visible distinguishing features. Definitely nerdy!

In the end it comes down to location. Marking our trip with stops in all the major valleys was one way to optimize our chances of seeing the many varieties of Rufous Antpitta and everything else because it was within this complex mountain geography that isolated bird populations, over time, evolved into new species.

Is this all hard to fathom? Probably. But it's a snapshot of my parents' extreme dedication to making the most of whatever birding opportunity was presented to us in South America.

For me, at ten years old, this was the least exciting exercise of the whole trip. It felt like we were seeing the same bird over and over again. But Mum and Dad loved it. In general, competitive world birders spend a lot of time thinking about this stuff.

(However, as I write this, Dad is peering over my shoulder and looking particularly smug. "I told you so!" And he had: I have gained *five* antpittas on my list because of him.)

Bosque Unchog is located in a tiny cloud forest valley, high up on the eastern slopes of the Carpish mountain range. It was a key birding site, with a handful of endemics. We wanted to be there before dawn and so were camping in the freezing cold. The zip didn't fully close on our tent, and we wore most of our clothes at night, including hats, scarves, and gloves.

It was an intense part of the trip: we were out all day, trekking the hills, only to return at dusk, exhausted, climb into our clothes, and zip up our sleeping bags. Although, to be fair, and in retrospect, reading about the legendary ornithologist Ted Parker's expedition to the area in 1974 makes me think we had it easy. His dawn-to-dusk hike up the mountain, ascending almost four thousand meters in fog; stinging, icy rain; and howling winds during his four-day search for the Golden-backed Mountain Tanager is either an epic tale of the power and passion of birding or a clear demonstration of how obsessed we all are, depending on your perspective.

On the first morning, our guide resisted the dawn start that had characterized so much of our South American trip. He insisted we enjoy a good breakfast as we took our places at a camp table set up beside the mountain road. Afterward, he insisted we take an hour to digest before we set off. It was all *very* confusing. Birdwatchers didn't sit down to breakfast, and they certainly didn't take time out to *digest*— they got up early and ate on the go, every single day.

We were on the verges of a small, steep-sided valley, almost barren apart from a little clump of stunted, elfin trees. As the sun rose, the shadowy side of the valley became bathed in bright light. Small orange fruits glowed in the branches of the copse below, and bang on time, the rare and spectacular Golden-backed Mountain Tanager arrived to take its breakfast too. Renowned for its tanagers, Bosque Unchog was a magnet for fans of this bird, the keenest up before the sun to catch a glimpse. But our guide was familiar with the habits of this particular tanager. He knew it would make an appearance around nine o'clock, with the warming rays of the sun, to feast on the red fruits. As my chilled bones began to thaw, the Golden-backed Mountain Tanager flew to the slope beside us to eat the red, coronet-shaped flowers that grew in our campsite clearing. In silence, we watched it eat, a small but spectacular bird with jet-black wings poking through a golden jacket and a sky-blue crown of feathers. When it had filled up, it hopped around the rocky clearing, probably to digest its food, before flying away.

It was a special moment, a unique event. That was the best day of the trip for Mum. Afterward, and toward the end of the trip, a couple of incidents signaled to both Dad and me that she was struggling; her tolerance for other people had run out.

We had just seen the rare and very beautiful White-bellied Cinclodes. A striking bird with pure white underparts and coppery brown wings had sat pirruping its song on a rock in the middle of a bog on a high mountain pass. Contented, we continued to scan the tufted grassy habitat to look for our next target bird, the Olivaceous Thornbill, but it was getting late, and now our guide, Alex, and our driver were keen to hit the road back to Lima. Dad had glimpsed the Olivaceous Thornbill while Mum and I were sixty meters away, inspecting the long grass for the sepia-toned bird. By the time we had joined him, the thornbill had flown.

Our driver was growing impatient. We were at 4,800 meters; soon it was going to be not only dark but also very cold and cloudy. But freezing temperatures and poor visibility were mere irritations to Mum.

The Olivaceous Thornbill is a tiny, mostly brown hummingbird, difficult to see as it skits between small pink flowers barely above ground level. For another hour we searched, Mum belligerent, insisting we stay until she was satisfied.

"This is needle-in-a-haystack territory, Helena. Let's go, please?" Dad moaned. He was right: the landscape was like Dartmoor, the ground covered in thick, heatherlike grass.

"You've seen it—of course you're ready to leave!" Mum snapped. Dad wasn't going to push it; he stood down while Mum and I continued to search. It was almost dark when we finally saw the Olivaceous Thornbill's bright green beard in the fading light.

"See?" Mum proclaimed, climbing back into the car. "If we'd left when you wanted, we wouldn't have seen it. Would we, Mya?"

I just shook my head slowly. It's not much fun scrabbling around in the dark while everyone else is waiting to go home, but Mum didn't care about that. She was elated. She could go from rage to pure joy in seconds: one minute, there was no thornbill and no possibility of ever leaving this increasingly eerie mountainside, and the next, she was practically buzzing with positive energy. I didn't mind this, so long as she was happy at the end of a birding day. It's when she doesn't get her bird, despite all our efforts, that a darker, longer shadow is cast on whoever is close by. Which is what happened a couple of days later.

We were driving toward the high Andean valley of Soraypampa, a traditional starting point for the trek toward the ancient Inca ruins of Machu Picchu. Mum and I were dozing in the back of the 4x4 when an Andean Tinamou—a heavily patterned brown-and-tan bird with chicken legs—flew across the road. Dad, ever vigilant, saw it, of course, but it was gone in a flash.

The tinamou had flown into the long dry grass by the side of the road. The driver pulled over and Dad woke us up. Mum was out of the car in seconds, moving fast into the fields while I tried to keep up, Alex all the while bellowing directions behind us. We searched for a long time, Mum carefully parting what felt like every blade of grass, in case the bird was somehow hiding there. But we neither heard nor glimpsed it.

"You should have woken us as soon as you saw it!" Mum shouted at Dad as we made our way back to Lima. "And you should have given us better directions," she snapped at Alex. "'In the long grass over there' isn't good enough."

Alex was a patient man, no doubt used to the obsessive temperaments of world birders, and he reassured her we would see more farther along the road, but Mum wasn't feeling reasonable, and there were no more tinamous to be had that day.

Tinamous are a family of birds that, to be honest, are not going to win any beauty contests. With their cryptic plumage and shy and retiring nature, I suspect some species actually wear invisibility cloaks, they are so seldom seen. When you are about the size and shape of a chicken, this is probably a very sensible strategy in order to avoid ending up as the next meal for man or beast. While they are restricted to South and Central America, they are closely related to the ostriches of Africa, the emus and cassowaries of Australasia, and even the extinct moas of New Zealand. They are our direct link to Gondwana, a supercontinent that existed 550 million years ago, which subsequently broke up into the continents we know today.

Mum's anger, once again, took a while to abate—three days exactly, which was when we finally caught sight of another Andean Tinamou. Her anger isn't everyday rage; she isn't an angry person by nature. Her circular thinking—in this instance, an inability to let go of her failure to see the tinamou—is a symptom of her bipolar disorder. When Mum becomes fixated on an idea—any idea, not just birds—she is unable to see it from any other perspective than rage.

By then, the Craig Family Harmony Index was very low.

I learned on this trip never to chase after a hummingbird. They are always much faster than you. It was while I was running up a hill after a Coppery Metaltail that I collapsed. This was partly due to altitude sickness and partly due to sprinting up a steep slope. Dad scooped me up and took me back to the tent where I was fed some chocolate. Where were the coca leaves? I wondered.

By the end of the trip, I had worn through my hiking boots; even the duct tape I had carefully wrapped around them refused to stick. I put them in a bin at the airport and arrived back in the UK in late August in my wellies. I was ready to eat some real food. It was hard being a vegetarian in the mountains of South America, and I had had my fill of eggs and rice.

On the way back, our flight broke for a day in Atlanta in the US, and we squeezed in thirty-five new birds before embarking on the final leg home.

While Dad and I felt invigorated by the trip (despite the fact that the Harpy Eagle had once more eluded us), Mum couldn't have managed another week on the road, even though she insisted she would have been happy to stay another six months. It was obvious she needed some of the routine and comforts of home.

Before we started on the Peru leg of the trip, Mum had hoped for three hundred new birds, which Dad believed was wildly optimistic, but we recorded just over one thousand in total. Many of these were birds that I had seen in Ecuador, Colombia, and Bolivia; the total number of *new* birds I saw in Peru was just over 360.

My world list was now over 2,900 birds. Given there are ten thousand species in the entire world, I was very excited.

It's hard to boil down the entire six months into a single overriding emotion, and I guess there's no need, but it was in the Bosque Unchog cloud forests, walking with Mum, that I came closest to capturing what was so special about those months in South America. What I loved most about our trip was getting to know my parents again. Before, and for so long, Ayesha had taken on the role of "parent." Mum and Dad had jobs that demanded more of them than the routine nine-to-five. They'd be there in the mornings and late evenings, but our daily contact was brief. And when Mum was no longer working, she had been either manic or depressed.

While it wasn't plain sailing for her in South America, she was there for me in a way she hadn't been for so long. I didn't want things to go back to the way they were before. I didn't want her to self-isolate with depression or become lost to us in her delusions. I had glimpsed more than birds in South America: I had found my parents.

6

The Dinosaur Bird

SOUTHERN CASSOWARY

The Southern Cassowary lives in the tropical rainforests of Indonesia; Papua New Guinea; and Queensland, Australia. It is a large flightless bird and a relation of the emu, ostrich, rhea, and kiwi. While encounters with the Southern Cassowary are rare, these birds come equipped with a fearsome reputation. Their triple-toed feet are powerful—featuring a lethal daggerlike claw of up to twelve centimeters on the inner toe—and they can execute a robust kick. The female is dominant and larger with a longer casque, bigger bill, and brighter-colored bare parts. It is the male who builds a nest on the forest floor, incubates the eggs, and raises the chicks alone.

Cassowaries forage for fallen fruit and can digest those which are toxic to other animals. As rainforest regenerators, they are able to distribute the seeds of over two hundred species in their droppings.

In fact, some rainforest fruit-tree seeds cannot germinate without first passing through the stomach of a cassowary.

I felt as if I had been away for years, and totally out of sync with normal life. Everything had changed, and yet nothing had changed. Most of my friends had never left the country before, and they weren't that interested in my tales of tropical jungles and glittering hummingbirds, and while I wasn't that invested in catching up with all the ins and outs of school gossip, I couldn't figure out exactly why I suddenly felt so separate from a group of people I had known pretty much all my life. Looking back, it's obvious my perspective of the world had changed, expanding in a way that couldn't be contained by village life, which, in turn, affected how much I could relate to others and they to me.

My life had flipped from a South American quest to sitting down alone in my room studying for my standard attainment tests, which were fast approaching. Very quickly, our extended holiday began to feel like a dream.

But I *was* excited about secondary school the following year: there would be new friends to make and adolescent milestones to meet.

At the same time, I began to feel like I was living a double life. A clock was ticking inside my head, counting down to our next birding break, when I would jump into the other part of my life, the part with the birds. While I was away, the countdown would start up again, calling me back. I wished there were less of a distinction between the two halves, but given that very few of my friends were interested in "Birdgirl," and I was even more reluctant to disclose my birding adventures, there wasn't much opportunity to bridge the divide.

Birding aside, my friendships were strong. I joined Girl Guides, where we hiked, camped, sailed, and canoed. Boy Scouts was more challenging and more fun, with its wood-chopping, fire-starting, caving, abseiling, and night-hiking focus. I still enjoyed sleepovers and the cinema with my friends, and maybe I should have tried harder to share more of my birding life with them, but I didn't. I never spoke about Mum's illness to anyone outside the family unit either, not until I was well into my teens. I didn't know or believe there was anyone who might understand what it was like to live with a mother whose mood swings were so erratic that I didn't know what kind of mum I'd be waking up to from one morning to the next.

I couldn't, in a million years, imagine a conversation in which I might tell a friend that I had caught Mum trying to flush her meds down the loo last night because Dad wasn't around at that moment to make sure she took them. What words would I use to explain that when Mum begins to feel manic—and it doesn't take much, just a night or two of sleep deprivation—she feels like abandoning her meds altogether? A couple of days without them, and she is determined never to put a single pill in her mouth again, because they will only send her to sleep, and Mum is *busy*.

Why's she so busy? What's she doing all day if she's not working? I imagined a friend asking. And my reply, idiotically vague, might be, *Oh, you know, just stuff on the computer. Half the night sometimes.*

Really? Doing what?

I have no idea.

Mum's spirits remained high after we returned, and for a couple of months, she, Dad, and I feasted on our trip, recalling each extraordinary bird as we drew up and compared our lists. Family, friends, and twitchers were regaled with the many highlights. The impact of the trip on our lives, despite the fact we were entering the winter months, was sustaining, especially for Dad. We had taken a long trip, seen extraordinary birds, and survived to tell the tale; he was already thinking of where we might go next, but by the end of the year, Mum was ill again.

She began to alternate between lying in her bed and lying on the sofa. She wasn't interested in me or what was going on around her; she seemed to be locked inside her own head. At times it was like sharing a house with a quiet, deeply unhappy ghost. For the first time in my memory, Dad, frustrated by Mum's lethargy and in an attempt to stay sane, was leaving the house to go birding locally on his own. And for the first time in my life, I was losing it with Mum.

I couldn't understand her: I was still thinking about South America, still basking in its afterglow. It had been an incredible trip, during which, and for long periods, Mum had seemed "normal." What was the point of going anywhere if all the benefits could be undone so quickly and easily? You get ill and then you get better, I reasoned. Why wasn't she? But then again, hadn't I felt weird coming home, finding everything the same, everyone talking about the same old things, following the same old routines? Maybe Mum, like me, was just finding it hard to readjust.

She has always been an unstoppable force, someone who prefers to move than to sit, to talk rather than ponder, and to throw her energy into helping those who are struggling. I became as frustrated as Dad when she disconnected from us, but deep down, he was still hopeful that she would get better. Her health had improved during the trip, and he would carry on trying to find solutions, seeking out ways to turn Mum back into the person she used to be. But I wasn't so sure. For the last three years, I had watched her bouncing between quiet despair and raging anger, and finally began to suspect that she might be like this forever. That while she might seem well, maybe even for long periods, eventually she would be sucked back into her sadness.

When she wasn't down, she was manic: reckless, irritable, and short-tempered, snapping at me if I was late for school or couldn't find my shoes. And when she shouted at me, I shouted back. I was growing tired of the usual excuses and claims that she would get "better." I wasn't an idiot. Maybe I wasn't angry so much with her, but with the idea that she might *always* be like this, not in control of her emotions or her moods. I felt abandoned, and this expressed itself in bigger and noisier arguments every day. If I was being honest with myself, I resented her for something that was out of her control.

During the winter months, while I was at school, Mum had a therapist from the mental health services unit visit her every morning to help her to leave the house because she was becoming agoraphobic. Mum needed professional help if she was to carry on traveling. I didn't know any of this, wasn't aware that she was no longer driving her car or going to the shops. The more detached she became, the angrier I got. To get a reaction out of her—to get any response at all—I would shout at her, and she would shout back.

I know now that these rows *were* a connection, of sorts. Even though we might have been fighting, we were at least communicating.

Mum had been seeing a care coordinator every couple of weeks, who would check in with her to see how she was managing her symptoms since being discharged from hospital the previous year, but then the service was cut. She wasn't seeing her most recent psychiatrist any longer since they had clashed over his diagnosis. He believed that she was merely depressed and needed more citalopram to help her cope. Our GP was initially against referring her to another psychiatrist; he couldn't see the point of it.

Mum's previous, unofficial bipolar diagnosis was the closest she had come to feeling understood. She was desperate for another opinion, and finally her GP relented and made a referral for Mum to see a new, private psychiatrist. When the appointment came through, she was instantly more hopeful, her mood suddenly lighter, but I didn't understand how this might change anything. Wouldn't we simply be given a new name for an old problem?

In December 2012, within the first session and without any bedside manner, the new psychiatrist reviewed all the evidence and confirmed my mother was suffering from bipolar disorder. He also explained that it *was* the SSRI—the citalopram—that had triggered the full-blown mania leading to Mum's hospitalization. Mum had been trying to manage her symptoms, unsuccessfully, since her early teens, and now she was going to get the help she needed: a new course of drugs, starting with lithium, followed by close monitoring to get the balance and combination of the medication right.

When they got home that night, bursting with news, I sensed a turning point for my mum. Not only did she finally have a definitive diagnosis, she also felt someone was really listening to her, for the first time.

"*Really* listening to me, Mya," she emphasized.

"And me," said Dad.

Mum laughed as she told the story, but I didn't think it was that funny. She had instructed Dad not to go into the consultation with her—she wanted to talk to the psychiatrist alone—and Dad was dispatched to walk around the block. Just as he set off, she phoned him, urging him to come back to the office; the doctor wanted to talk to him, too, to hear his thoughts and understand his experiences of caring for Mum. When Dad, in tears, explained his efforts to keep her alive, safe, and comfortable, the doctor turned to Mum.

"It's time you stopped fighting Chris, Helena. If you don't, he'll be the next member of your family to end up in hospital. It's time you returned the favor and begin to care for him the way he cares about you."

My parents' jaws dropped.

"It's not all about you," the doctor went on. "It's about Mya and Chris as well. You're ill, but you still have control over how you treat others."

Mum, to her credit, had felt empowered, if anything, by his direct manner. It was true, she had reasoned: Hadn't she lashed out at Dad when there was no way on earth he was responsible for her sadness or her mania? Hadn't he taken the brunt for the birds she had missed in South America?

It was with a twinkle in his eye that Dad told me the psychiatrist's parting shot to Mum: "Chris is the only person keeping you out of hospital, so it's time to show him a little appreciation."

"I do appreciate him, you know," Mum told me. "Just in my own way."

For the first time in her life, she hadn't been made to feel, as so often had happened in the past, that her mood swings were part of her personality. My parents were hopeful, at last, that with some help, she might learn to enjoy life more consistently.

I was relieved they seemed happy—on the same page about her treatment—and that the future looked a little brighter. The doctor had been right about Mum's willfulness, her unreasonable expecta-tions of Dad, her snappiness toward me, but bipolar disorder is a vi-cious illness, and it made her selfish and isolated at times. It is also a home-wrecker; while my parents managed to hold their relationship together, it's no surprise that practically every other sufferer in my mum's circle from hospital had split up from their partner. And if her relationship with me was anything to go by since we had returned from South America, I imagined they didn't get on much better with their children.

However long it took Mum to get diagnosed, she believes it would be much harder to get that diagnosis today given the reduction in men-tal health services in the NHS. When she first became ill in 2010, there was a local hospital with a separate mental health services building. She was offered psychotherapy for six months after being discharged from hospital, and a course of family therapy. Today, the same level of support is not available, and often, bipolar disorder is still confused with a host of other mental illnesses.

My *nanu* was the one person who wasn't happy about the diagnosis. Why did Mum have to have such a shameful label? she wondered. What good would it do for all our relatives to know she was ill? She insisted Mum keep the news to herself, as there were cousins and nieces whose chances of marriage might be badly affected if it was known there was mental illness in the family. "No one's even having arranged marriages anymore," Mum objected. But my *nanu* is stubborn, and she shut down the conversation. She wanted to bury it.

The denial of mental illness is strong in my maternal line. It sounds strange, but it was almost a joke in the family: Mum wasn't mentally ill—she just had an anger-management issue. At least my *nanu* had moved on from believing that Mum was possessed by a jinn; she was convinced the *tabiz* had worked! I had grown up hearing stories of jinn and shrugged them off just as Mum did. These were the ramblings of my elders, I had decided. They had nothing to do with me and Ayesha, or Mum and Dad.

While Mum didn't let her family's opinions knock her confidence, now that she had her valuable diagnosis, I knew enough to understand the more sinister side effects of mental health denial in our communities. If an issue was so taboo, why would anyone in the family ever seek help if they, too, became ill?

We could have benefited from their help. If there hadn't been such a backlash against mental illness, they might have stepped up a little more. The stigma is down to superstition and the idea that a "good" family is a family free of any gossip around reputation. But it's a great shame, especially today, when mental illness is discussed more freely than ever before.

The rift in my relationship with Mum was upsetting for her and for Dad. On the advice of her GP, she and I were to begin a series of group therapy sessions run by local social workers. Initially, I hated the idea.

I didn't want to sit around and discuss my "feelings" with a bunch of strangers. I was ten years old; it was an awful idea.

"But they're not about talking," Mum explained. "Not at all. We'll just be hanging out and doing stuff together."

In truth, Mum was equally reluctant to go. On our very first session, we got lost on the way there and discussed turning the car around and forgetting the whole thing.

But somehow we arrived, and with ten other "troubled" families with kids of all ages, we set about "doing stuff together," which involved a lot of craft activities.

"Fill this jar with water and add some glitter, Mya," I was instructed. I filled the jar and screwed on the lid. "Now, when you're feeling angry, flip it over." I flipped it over, and the glitter cascaded down. I felt like laughing, which maybe was the point: better to laugh than smash the jar against a wall, I guessed.

If I enjoyed these sessions at all, it was because I had Mum's full attention. She wasn't working late into the night or dashing around, trying to right the world, and she wasn't lying in bed with the curtains closed. It wasn't quite the same as walking and talking in the cloud forests of the Andes, putting the world to rights, but it was still good. We painted masks together and ate popcorn around a campfire. A year earlier, during one of Mum's "stay in bed" periods, Dad had taught me to ride a bike. And now, for the first time, Mum and I rode our bikes together through the woods. But the activities weren't important in themselves: it was about the time we spent together. And for that reason alone, I felt my anger begin to ebb away and hope inch in.

Six months had passed since Mum was diagnosed, but in reality she was just beginning her journey to "recovery." At this point she was only on lithium, a mood stabilizer that would hopefully help with mania in particular. It would take a long time to balance the mix of drugs—new pills would be added one by one while Mum was monitored for her reactions.

Mum seemed "different" since starting the drugs: she was more functional, her moods had stabilized, and she was less likely to fly off the handle. But there were also days and weeks when she was exactly the way she had been before: staying up late into the night, researching online whatever new fixation had taken her interest.

It was in these early days of adjustment that my nephew, Lucas, was born while Mum held Ayesha's hand. I was anxious at the time about the impact his birth would have on my own parents' anxiety levels, but I needn't have worried. They were true to their word: when Ayesha announced her pregnancy, she had made an active choice, and they would be there to support her if she needed them. This time around, Ayesha was more secure, and she was a mother already. There wasn't the same scrabble to get things ready for her, to sort out her accommodation, to make sure she had enough money to fill the fridge. Ayesha felt calm enough to paint rainbow murals on her new baby's bedroom walls.

Dad was itching to lock down a destination for my long summer holidays, and so my parents began to scour species lists and birdwatching websites, before finally committing to Queensland, Australia, for the six weeks. Traveling around Oz by camper van seemed like the most cost-effective way to make the trip, as well as turning it into an adventure, and Mum was in charge of booking the van and the first week's campsite accommodation. It was only as our departure date inched closer that it became evident she hadn't organized very much at all. Dad had taken the lion's share of planning our South American trip, and Mum felt Australia was her turn. She was on a mission. With two weeks to go, we still had no vehicle. Dad wasn't happy, but Mum on a mission was not to be messed with and he didn't dare to intervene.

Mum's Bangla mindset of leaving everything to the last minute and beyond clashed with Dad's desire to have it all sorted beforehand, allowing extra time to deal with any problems, delays, or mishaps. If we're flying, Dad prefers to arrive several hours before takeoff, providing plenty of time to check in and get through customs, while Mum is happy to assume there will be no issues with the journey to the airport or at security and leave everything to the last minute. To get her out of the house, Dad often threatens, "If you're late, they won't hold the flight for you, you know." I suspect, however, that Mum doesn't actually believe they would dare take off without her.

When she did eventually get around to it, Mum found loads of companies that hire out camper vans, but the Queensland Music Festival had cleared them out. The week before we left, she was still up at midnight most nights, frantically ringing car-hire companies, begging them for "anything."

"And the campsites?" asked Dad, a little desperately. "Tell me you've booked at least a few."

"It will be fine," Mum repeated. "And we can book along the way, can't we?"

We arrived in Brisbane to find festivalgoers—in camper vans— everywhere. I didn't mind that our van was very small (the only model available). I got to sleep in a tiny canopy bed, lying just centimeters above the heads of my parents in their bed. It was also rather basic, so no showers for the first leg of our trip, which suited me fine. (I didn't think too much about the fact we'd have no flushing loo.) My Kindle had been loaded, the Beastie Boys were shoved into the CD drive, and we were on our way.

Our first major stop was on the Queensland–New South Wales border. Lamington National Park is a vast reserve famous for its waterfalls, subtropical rainforest, and wildlife. Lying on the Lamington Plateau in the McPherson Range amid the remnants of an ancient volcano, it is a dream destination for birdwatchers.

We arrived late to our first successfully booked campsite and went straight to bed, only to wake up in the middle of the night because it was absolutely freezing. I started complaining: Wasn't Australia meant to be hot? Wasn't our journey from Brisbane gloriously sunny? But the plummeting temperature had taken my parents by surprise too. Experience pays off, however, and Mum has never knowingly underpacked: we had a good stock of thermals, so we pulled them on and slid back into bed.

Early the next morning, it must have been around six, I wondered if the sun would ever come out from behind the mountain peaks in the distance. We kept our thermals on for what promised to be a rather cold and wet first day of birding. The weather had a distinctly English feel to it; we might have been going for a damp woodland walk in the Lake District. The birds, on the other hand, had a distinctly un-English feel to them! We strung our binoculars around our necks and made our way into the rainforest. While Lamington is a busy park, it was still very early and very quiet. I was shivering a bit and still half asleep when we spotted our first bird. The Brown Cuckoo-dove was a disappointing specimen, however; it looked like any old city pigeon, and we had plenty of those at home.

"Was that Mya's three thousandth bird?" Mum asked. This perked me up a bit, snapping me out of my gloom about the weather and the lack of gorgeous birds. I peered at this mousy-colored specimen: Was this to be my milestone bird?

"No," Dad replied. "She's almost there." My excitement ebbed, but at least now I was focused, the chill in my bones easing as the sun finally appeared.

By this point, around nine o'clock, it was warmer, and we began to remove a few layers. Australia remained attached to Antarctica until around seventy million years ago, and the ancient rainforest in Lamington was still home to Antarctic beech trees, adding to its general air of timelessness. I felt as though I were in Jurassic Park, and while there were no dinosaurs (apart from the birds!), the forest had an uncanny air about it, clouds hovering in and over the tall beeches. Our target for the morning was the Regent Bowerbird, a beautiful milestone bird.

In common with other members of the bowerbird family, the male Regent Bowerbird builds a bower—an ornate avenue lined with sticks and decorated with berries, snail shells, seeds, and leaves. Sadly, these days it is also likely to include pieces of plastic, with a particular penchant for anything blue. The birds even go so far as to create a green or blue "saliva paint" in their mouths, which they apply to the walls of their bower with leaf "paintbrushes." They are one of only a small number of bird species that use tools. When the male is not busy attending to the multiple females (they are polygamous) visiting the bower, it is off raiding any decorations that take its fancy from other bowers in the area.

It is one of the iconic birds of Lamington, its habitat restricted to southeast Queensland and northeast New South Wales. We followed a path until we came to a clearing, with tall trees whispering to us that their secrets were there to behold, if only we looked.

Out of the corner of my eye, a flash of gold; I turned my face to the sun and drew in a sharp breath. There, perched on a high branch, basking and radiant, was one of the most spectacular birds I have ever seen. Any birdwatcher will tell you that there are some birds that were invented just to be stared at, and we were looking at one of them.

The Regent Bowerbird is entirely deserving of its name: it is *regal* from beak to claw, from its glossy black plumage and cape and crown of molten gold to its warm, liquid yellow eyes. And there was more than one; the bowerbirds glowed in the sunlight, as though the ancient beech tree on which they perched was growing gilded birds.

We watched, speechless and thoroughly enchanted, until something alerted the birds and they took off, spreading bright yellow wings, a treasure chest in the sky.

"*That's* her three thousandth bird," said Dad, grinning madly and pulling me into a hug.

Three thousand bird species is a lot, and for the first time in my life, I felt like a world birder. I needed a minute to think about what it meant. Realistically, to achieve this sort of tally, one needs to have visited a number of continents. Every thousand birds is a big deal for any birder because, let's face it, how many times are you going to reach such a landmark number? Of the 10,752 species in the world, only a dozen or so people have seen more than nine thousand, with forty people having seen over eight thousand. World birding is very much a personal journey—there is no official agency to record the greatest world birders and how many birds they have identified, but you are free to register and share your sightings on various internet sites. I personally have always used the BUBO list.

Even though I was just eleven years old, I recognized this moment as a turning point. It wasn't just a number; to me it represented all the fantastic birds I had seen over the years and how much effort I had put into tracking them down. But it was more than that; it was also the story of my family journey. The birds of South America were entwined with memories of our six-month "time-out"; those of Ecuador and Ghana would highlight moments of frustration for Mum as well as joy.

Of course, if the earlier brown bird had been the one to take me to my milestone, it would have become a "special" bird to me; it's true to say that all the birds on my list have something unique about them, something lovable, but I was a child and secretly thrilled that my landmark bird had been so pretty.

. . .

As we traveled through Queensland, heading north toward Cape York, my English brain struggled to rationalize the vast spaces we were covering. The Beastie Boys sang about how there were too many rappers and not enough MCs in the world as we plowed through the endless Australian outback. We drove for days, and yet we were still in Queensland! I had to get my head around the idea that while only a fifth of Australia, Queensland is easily seven times the size of the UK.

Mum was very sleepy in those early days, as she was still adjusting to the new drugs. She would doze in the passenger seat of the camper van while Dad drove *and* map-read, but this was by no means our biggest issue. Mum has since wondered why she thought she was capable of organizing any part of this trip on her own, but overarching ambition is a classic symptom of mania.

She had not booked nearly enough campsites, and we discovered that "booking them along the way" wasn't as simple as it sounded. The stress ramped up as Mum pleaded over the phone with camp officials to let us in. Some nights we just parked up on the side of the road and slept there, but this was a risk, as Australia has strict laws about where you can camp, and heavy fines if you break them.

Mum had also misjudged the distances between campsites. Once or twice, after a day's birdwatching, she would give Dad directions, we'd turn up very late and either get turned away or get an earful about their closing times and respecting their rules. Dad then had to drive on to try to find another site. I didn't think much of these journeys, wedged between my warring parents in a small metal box. I couldn't even sit in the back, because I would miss the birds. Luckily I had my iPod and headphones and *Now That's What I Call Music! 84*—forty-three tracks of distraction, interspersed with albums by Beyoncé and P!nk. Also, luckily, the stress and anxiety as we searched for a new campsite, sometimes in vain, evaporated once we were settled in.

Mum hates driving abroad, but eventually Dad insisted she occasionally take the wheel. These were generally white-knuckle affairs, and I wasn't convinced it was such a great idea, given that Mum froze every time a road train, trucks pulling multiple trailers in a long chain, shuddered by.

Part of our morning ritual involved Dad asking Mum to start calling campsites for that evening. She often wouldn't get around to it until lunchtime, and the whole panic would begin again: a full campsite, too late for the next, and then, in despair, pulling up on the side of the road to sleep, praying that the police weren't going to hammer down our door the moment our heads hit the pillows.

Part of Mum's reluctance to get on the phone first thing was her fixation with the landscape. She and I zoned out on these long hours of driving, watching the land unfurl. Compared with the English countryside, where even the most rural landscapes have been shaped by people, the emptiness of the outback was mesmerizing. If we ever came across animals, they seemed so tiny, dwarfed by the never-ending wilderness. While every inch of the UK is teeming with human, animal, or insect life, these sprawling spaces regularly sent my brain into a spin.

The main signs of animal life in this rugged environment were the corpses of kangaroos and wallabies strewn over the hundreds of kilometers of roads we traveled, all of them victims of moving vehicles. I took the wallaby deaths particularly hard, as these animals had been very tame around our first campsite at Lamington National Park. I had been happy to stare at them for hours, and now, a week later, I was watching Black Kites hovering over these scenes of carnage, picking apart the remains, delighted by their abundant feast. And it wasn't as if the roads were full of traffic. I was puzzled by the number of deaths until early one morning, as we made our way down another long stretch of boiling tarmac, we saw a van neither slow down nor swerve to avoid the wallaby in its path. It hit the animal hard, killing it on impact, and sailed on. I wondered if it was a sport for these people.

We were winding our way north toward Cairns, our journey taking in coastline, rainforest, outback, and empty roads.

In Cairns we picked up our second camper van, a 4x4 geared to cope with the rough pitted tracks on our journey up to Cape York. This camper van was even smaller than the first; it was more like a big car, and the CD player would accept only one disc of *100 Hits: Driving Rock*, which regularly got stuck on "Road Rage" by Catatonia, weirdly appropriate.

But none of this really mattered as we headed for Mission Beach. We were focused on our target bird and one of my favorite species: the prehistoric Southern Cassowary. Two meters tall, it looks like a dinosaur crossed with a turkey. With a blue face and Mohawk-like red wattle for a cap, this flightless bird has dagger-sharp claws on its feet and can kill a person with a single kick.

Mission Beach, two hours from Cairns and basking at the edges of the blue waters of the Coral Sea, flanked by World Heritage tropical rainforest, is home to one of the highest concentrations of the vulnerable Southern Cassowary in Australia. I was desperate to see one, but they can be difficult to track down. In general, the bigger a bird is, the fewer there are of them. Bigger birds need more food, and any given habitat can sustain only a certain quantity of its wildlife; any more and someone is going to starve. Animals are good at self-regulation, however. If there are too many in terms of territory, food, or resources, their numbers fall. In the natural world, the cycle of life is in balance; it's only when *unnatural* outside forces interfere—namely human beings—that food abundance is affected and species become endangered. There are many stories of the Southern Cassowary wandering into urban environments in search of food, no longer able to find enough in its diminishing natural habitat.

While a bird strolling across the road at Mission Beach was my first sighting, my most memorable experience was in the car park in Mount Hypipamee National Park. No sooner had we slammed our camper van's doors shut than a tourist began to yell, "Cassowary!" Strutting around the cars and vans was the six-foot-tall bird, with two small and very cute chicks. Their father, however, was a huge specimen. I took a step back—a huge, *scary* specimen. He had an abundant glossy black plumage, like a fur cape draped over his back and shoulders, clashing dramatically with the bright flash of his long blue neck.

The tourists were snapping away with their phones while the male drew closer. "It's getting agitated," warned Dad. "Move back!" The cassowary did seem agitated, taking step after careful step on long legs toward its snap-happy audience. The big bird looked like he was about to leap—Mum pulled me behind her. At that moment, the chicks began to call; they had had enough of this crowd. *Let's go, already,* they seemed to be squawking, and with that, all three birds turned tail and darted into the forest.

Back in the camper van, recovering from the adrenaline rush of fear, I was delighted we could add the Southern Cassowary to our day list, and we didn't even have to leave the car park!

No one in the family was missing the presence of a guide at this point. We were enjoying having to rely on our own research and the online twitching community. We went wherever our leads took us, and while we weren't going to find any of the secret spots, which good guides are famed for, this trip, rows aside, had a freer vibe. In the evenings we sat outside and compared our bird lists, relishing the new additions. The Craig Family Harmony Index was at its peak during these sessions, happier outside the van than in.

We left Mission Beach and headed toward the Mount Carbine

caravan park in Far North Queensland. The bird we were after was the Tawny Frogmouth. Owls, frogmouths, and potoos are very reliable and often roost in the same place every day, making them relatively easy to see if you know where to look. The twitching hotline had alerted us to a Tawny Frogmouth roosting in a tree in the middle of a campsite.

We were keen to pursue it, but we couldn't just wander in. Mum's persuasive powers came to nothing when she went to speak to the camp officials. Dad joked that she was losing her touch, but Mum didn't think he was very funny, and anyway she was now looking at me. "You're just a shy eleven-year-old girl, Mya, they'll *love* you. Soulful eyes, okay?"

She gave me a little shove, and off I went, mortified. But...I also really wanted to see the Tawny Frogmouth. In the end, I didn't have to say much more than, "Please let us in—we've come a long way and it's a very special bird." Obviously, they felt a little sorry for the tongue-tied kid with the pushy parents and waved us through the gates toward the tents and camper vans parked in the midst of towering, leafless trees.

Accompanied by the site officials, we began to inspect the bare branches with our binoculars.

"There!" said Dad, pointing into the uppermost branches of a tree directly over a camper's pitch.

It was a brown bird in a brown tree; it's amazing he spotted it at all. With its squashed head, the Tawny Frogmouth looks like it gave up trying to become an owl at some point in its evolution. We saw it only because its beady eyes glinted a bright orange in the sun. Its brown speckled plumage was the perfect camouflage. But then the strangest thing happened: the bird opened its beak, revealing a fleshy lime-green interior, and its entire face split in half, hence *frogmouth*. It was when we all started laughing that the camp officials decided they'd had enough of this weird family and escorted us off their land.

• • •

As we moved farther north, I came face-to-face with the extreme poverty endured by the Aboriginal communities who live on the reserves. In Lockhart River, a coastal Aboriginal town on the Cape York Peninsula, I watched children my own age in rags playing in the streets. So far, we had not seen an Aboriginal face in any of the towns we had visited.

When I visited Bangladesh in December 2006—I was four years old—we traveled around Sylhet District, visiting my *nanabhai*'s village. We watched the pilgrims congregate at the Shah Jalal mausoleum and noted the children begging on the streets of Dhaka. Although there was deprivation, the kids were dressed well and wore shoes. Maybe I had been sheltered from the worst of Bangladesh's poverty, the child prostitution hidden away as it is in many parts of the world, but what I *saw* did not look catastrophically desperate. The shanty towns had running water and their inhabitants carried jute bags filled with food.

But this was Australia! A developed country with a high standard of living. Brisbane was clean and felt moneyed, a glimmering city of skyscrapers. Cairns was a lively place full of hip tourists and pricey restaurants. My parents continually discussed the shockingly high cost of food. Everything I saw and experienced of Australia pointed to the comfortable lifestyles of its inhabitants.

We all have preconceptions about the places we visit, and even though I was just eleven years old, I had my own. I believed parts of Asia such as Malaysia and Hong Kong were relatively prosperous. South Asia was poor but people had clothes, food, and water. Africa was poor and people did not always have basic amenities. North America was rich. South America was a mixed picture: some countries were well off, and in others subsistence farming was the norm. Europe, all of it, I deemed wealthy, where even basic living standards provided food, shelter, and amenities.

I also considered Australia wealthy. And this was my experience until we reached Lockhart River. Here I was in a rich country full of White people, where everything was really expensive compared with the UK, and now, in this town of Indigenous people, everyone was wearing rags and walking around barefoot. The contrast was shocking.

How could such pockets of poverty exist, given the metropolis of Brisbane? I had witnessed poverty at home. I was used to seeing home-less Bristolians, but this was explicitly racialized poverty. I was looking at an entire community of a single ethnicity without the resources that were available to everyone else.

Just after we returned to the UK, we watched *Australia with Simon Reeve*, in which the documentary filmmaker travels to Cape York and talks with some of the Indigenous population. He reveals a situation far worse than the mere snapshot I had witnessed.

It was clear that the plight of the Aboriginal peoples in Cape York was down to racism, later confirmed when I began to learn about the racist legacies of European colonialism.

My route to challenging racism and becoming an activist stemmed from my anger and sense of injustice. What made me furious in Australia was not just the poverty but the comparative poverty. The guilt and shame I felt only intensified as I grew older.

Cape York is an important destination for birders: the largest tract of remaining tropical rainforest, it is also home to a number of species found nowhere else in Australia. One of these, the Golden-shouldered Parrot, a black-capped blue bird with golden shoulder patches, inhab-its a small area at the base of the peninsula. Its very specific require-ments of a certain seed to eat and termite mounds of just the right size and shape to nest in have made it a rather rare creature, with only around three hundred breeding pairs. We camped overnight at Artemis Station, which has been carrying out pioneering conservation work to protect the parrots since the 1970s, and were lucky enough to catch a couple the next morning. I could have sworn they were grinning at us, their beaks shaped like smiling mouths.

It became very obvious, very quickly, why we needed the 4x4 in Cape York: the roads were just tracts of packed dirt and craters. As we drove north, we regularly encountered road trains. They thundered toward us, churning up the red dust of the road, completely obliterating the sun, and reducing visibility to zero. There was nothing Dad could do but pull over and wait for the cloud to settle and the road ahead to reemerge from the choking dust. There were few other people around, and the farther north we went, the more tropical the climate became. It was tough going, and Dad was struggling to keep the car upright and moving forward. By the time we reached our very basic campsite, it was no surprise to find it populated with the toughest of travelers. They looked as if they were equipped to survive for months off-grid if necessary—in fact, some of them may well have done so already.

Early the next morning, we were following the rough paths of the rainforest in Kutini-Payamu (Iron Range) National Park, looking for our target, the Magnificent Riflebird, in the only area of Australia where it occurs. It was very quiet and dark, the foliage so thick that we had to take it slow. Eventually, at the edge of the forest, we heard a single bird begin to call. Tracking the noise, we got closer and closer to where it seemed to be coming from.

Riflebirds tend to sit high up in very tall trees, so our binoculars were aimed toward the sky; it was hard to walk and to look up at the same time, but we pushed through, calling, pausing, and following. This chase went on for a couple of hours, but the key to successful birding, as ever, is patience, so we waited, staring into the upper branches of the towering jungle trees. And then we saw it. I remember feeling a twist of disappointment. The riflebirds are classified as birds of paradise, a species usually very brightly colored, exactly the sort of birds you might find in an Attenborough documentary. As well as their rainbow plumage, they also sing beautifully and have killer dance moves, but the Magnificent Riflebird looked more like a blackbird. It would have been impossible to identify in the trees if it weren't for its iridescent turquoise triangular breast shield, which saved the day, catching the sunlight as it took to the air.

We picked up our third camper van back in Cairns. It was much larger and definitely more luxurious than the first two. This was more like it: space to breathe; the walls lined with sofas, which unfolded into a full double bed. But...this van, unlike the second, couldn't be driven off paved roads, and our next destination, Lawn Hill National Park (now Boodjamulla National Park), was a two-hundred-kilometer round trip on unsealed roads. We were going to take our chances.

After days of driving along endless roads in the dusty orange outback, Dad was exhausted, so Lawn Hill National Park, an oasis of green, was a joyful retreat. Rough tracks negotiated and thankfully crossed with ease, we parked at the edge of a topaz-blue river, which cut through a canyon of craggy rocks. Trees lined the banks of the river and, farther along, a waterfall. However, we were looking for the Purple-crowned Fairywren, which wasn't to be found in this lush landscape. Leaving the van by the river, we walked toward the dry bushes and rocks of the barren outback. This time there would be no waiting, because perched on a low stump was a male fairywren, strutting his stuff for all to admire.

A mostly brown bird, he was lucky enough to have a long and shimmering bright blue tail, cocked for all the world to see, the same blue as the river back in the park. A beautiful purple crown sat atop his head, and a black bandanna wound around his eyes. He was gorgeous. My excitement had nothing to do with sighting a *rare* bird: just watching this miniature marvel was enough. He came to rest on a gray boulder, where he was soon joined by his friends. They proceeded to flit around the dusty outcrop of rocks while we stood there, mouths open. They didn't seem to care about us one bit, and we were in no hurry to go off and start looking for other birds; for a while, both birds and humans enjoyed each other's company.

Mum was well aware that the trip had been tough on Dad, but on our return home, she nonetheless wished that it had been longer. This feeling was already part of her pattern: while she was away, her desire to stay in the small cocoon of our family was a strong one. She had missed Ayesha and Laila, of course, and the new baby, Lucas, but there was something about cooking, sleeping, and birdwatching together that strongly appealed to her. For a short while, she was taken out of her own life and placed into one where the only concerns were where we were going to bed down that night and which birds we would chase the next day. Mum could be manic while we were away, but the pendulum of her bipolar disorder rarely swung toward depression on these trips.

The activity sessions in the woods with Mum a year earlier had brought us closer; I wasn't so angry these days, but she was still worried I wasn't talking to anyone about the impact of her illness on my own mental health. It was after I began secondary school in the autumn of 2013 that we were referred to the Child and Adolescent Mental Health Services (CAMHS) for family therapy sessions. These were presented to me as sit-down sessions where I would be given an opportunity to discuss with a counselor my feelings about Mum's illness and how it had affected our relationship.

The thought of it alone made me feel furious. I didn't want to talk to a stranger about my mum. If I had felt closer to her recently, her insistence that these sessions would be good for the whole family set me right back and I was angry with her once again. To appease my parents, I went to one session. A stark white room, silent except for the gentle prodding from the counselor to "open up," Mum and Dad *staring* at me like they were waiting for me to break into song—it was, as I'd suspected, awful. I had loads of feelings, of course I did, but I was very young and lacked the emotional maturity to put these feelings into words. I was also still living with Mum's mental health issues; it wasn't as if she was better now, or that we were out the other side. I refused to go again.

With hindsight, there was nothing intrinsically wrong with the sessions or the counselor, but I wasn't ever going to be happy with the spotlight on *my* feelings, in a sterile room under strip lighting, while a stranger talked to me about my mum. Coaxing my feelings to the surface would require a different language.

"You know what, Mya," Mum said one late-autumn morning as I was getting ready to leave the house. "I think some *poetry* sessions are just what you need."

(This might seem like a lot of therapy in a short space of time, and that's exactly what it felt like.)

"Poetry?" I laughed, but she was being serious.

Mum wouldn't let it go. She was convinced I needed to express my feelings or they would rot away inside me and make me ill, but CAMHS had been a disaster as far as I was concerned. "Will you and Dad have to be there?" I asked.

"Of course not. The family format doesn't work for you. That's obvious." What was obvious was that Mum wished that I had spilled my guts to the counselor. Well, that was never going to happen, but to pacify her, I agreed to give her latest idea a go.

And so I began a series of poetry therapy sessions, which involved sitting with Ita, family friend and poet, while she tried to coax a few thoughts from me that we would then mold into a poem. It felt very childish, staged, and uncomfortable, and to begin with, I didn't enjoy anything about the sessions at all. Initially, I would come away feeling quite smug because I hadn't revealed anything of myself, but gradually, without realizing it, I began to open up. Within a couple of weeks, something changed. I was asked specific questions rather than those of the vague "How do you feel when your mum is sad?" variety. We drilled down to certain events, like when I first visited Mum in hospital. I found myself talking about the nurse "guarding" the room, the smell on the ward, and the fact that Mum was trying too hard to be Mum.

If it was a surprise to Ita that the words suddenly began to flow, it was even more staggering to me. I began to feel lighter after these sessions because, very simply, it was the first time I had talked to anyone about my life.

As part of getting better
We tried a family hug in the kitchen fairly recently
It was the only time I
Saw my daddy cry
It was full weeping
But we were trying
So that's what matters

7

Roots

SPOON-BILLED SANDPIPER

The Spoon-billed Sandpiper is a small, critically endangered wader that breeds in northeastern Russia and winters in South Asia. In 2016, the global population was estimated to be around two hundred pairs. The main threats to its survival are habitat loss due to climate breakdown on its breeding grounds, loss of tidal mudflats due to reclamation on its migration route, and hunting within both its migratory and wintering ranges.

This sandpiper's most distinctive feature is its black spatulate bill, from which it gets its name. It feeds in a distinctive style, sweeping its bill from side to side as it walks, sifting through silty waters for invertebrates to feast on.

A critical part of the conservation program to save this bird has been the education and provision of alternative livelihoods for bird hunters in the wintering grounds in Myanmar and Bangladesh, where the Spoon-billed Sandpiper is a victim of "bycatch" in the nets targeting larger waders for subsistence hunting.

The final entry on Mum's MECH list reads: *Jan. 2014. Mya concludes poetry therapy.* The list had been made for our family therapy sessions, and when they came to an abrupt end, so did the function of the list. Three lines above this entry is a more significant one: *Sept. 2013. Helena retires from work.* She hadn't returned to the solicitors' firm since she came out of hospital, but now it was official. And Mum was still too ill to think about getting another job.

Toward the end of January 2014, I launched my blog. It felt like a fairly easy decision at the time, and purely for my own enjoyment: to chat to the internet and my fellow birders about the birds I had seen and where I had seen them, and to generally share the "joy" of standing around in the rain, hoping to spot a rare migrating duck, goose, or wader.

When I was just eight years old, I had come across a cartoon superhero named Birdgirl, and while I lacked her silver wings (and yellow helmet), I identified with the name; I called my blog *Birdgirl.*

Swarovski Optik, makers of precision binoculars, had just featured a blog post about my having seen three thousand birds. My school newsletter picked it up, and then the *Chew Valley Gazette* ran it, and finally the national press featured it. My blog's popularity and growth took me by surprise, and I soon found myself with a reasonable platform, one that would quickly take on environmental and conservation issues. I began to blog about deforestation in South America—birds were becoming endangered because of loss of habitat there and all over the world. I was just as invested in the survival of the species as I was in watching them. In late December 2014, I wrote a blog post about the recent Sundarbans oil spill and went on to successfully campaign for awareness and funds, raising over $35,000. An ecological disaster, the spill had spread over 140 square kilometers, endangering the rare Irrawaddy dolphin as well as the plentiful kingfishers and egrets that visited this World Heritage Site. The spill hadn't happened in the sea, but in the mangrove forest, making it very difficult to clean up. I wrote an article that was placed in the American Birding Association's magazine, a publication with a very wide reach, and attached my campaign to actor Mark Ruffalo's activist group, Water Defense, because they are specialists in cleaning oil off the surface of water. With the magazine's readership of millions, the campaign hit its target the day after the article had been published.

All of this shined a spotlight on my work, and I found a wider audience for my voice. People seemed to like what I was saying and campaigning for. I began the transition from writing online to becoming more active in real life. When Bristol became the UK's first ever European Green Capital in 2015, the mayor put me forward to become an ambassador. I shared ambassador duties with Hugh Fearnley-Whittingstall, Kevin McCloud, and Simon King—our responsibilities were to promote the green capital, which I did on my growing social media channel.

Now I was meeting local nature and community wildlife groups to talk about world birding and biodiversity loss. I moved on to conservation groups and crafted my message as I learned. I spoke to the British Trust for Ornithology, and to the Chew Valley chapter of Avon Wildlife Trust, pressing home my message about declining habitats.

In January 2015, the BBC Natural History Unit filmed me while I spoke to the Wildfowl & Wetlands Trust about a project close to my heart: the Spoon-billed Sandpiper.

Later that year, I would address the Bristol climate change rally to mark the Paris Climate Conference, where participating nations gathered to combat global climate change. I felt energized and exuberant afterward—more than anything, the protest had felt empowering. This was three years before Greta Thunberg burst onto the scene; her message, School Strike for Climate, would mobilize millions of students the world over to become vocal activists for climate change.

To support the rally, I wrote a blog post, "How Teenagers Can Help Save the Planet," which led to my inclusion on the list of Bristol's most influential young people: twenty-four people under age twenty-four. I was becoming very recognizable, visually and by name; it felt like an organic and natural development.

I was offered a monthly column in the local *Chew Valley Gazette*, Birdgirl's Tails, about my adventures with birds and my thoughts on environmental issues.

I talked and talked with whichever group would have me. My message felt urgent; I was discussing the incredible birdlife I had witnessed and also the effects of habitat erosion caused either by climate change or by humanity, and I wanted to use whatever means were at hand to campaign for awareness. When I was thirteen, I met with my local MP, Jacob Rees-Mogg, to raise my concerns about hunting with dogs. We didn't agree on much, but he told me that I should think about going into politics, and in his way, he gave me the confidence to meet with and talk to people in power.

I was naturally shy, so there was a period of adjustment as my confidence grew, a thrilling and terrifying training ground. More than once, my meticulously prepared slide show would run into problems minutes before I was due onstage. But I muddled through.

It was an exciting period; social media was alive with talk of climate change, and I felt at the center of a new and growing movement. I had no grand plan, and I wasn't being directed by my parents or anyone else. I was simply following my passion.

. . .

Somehow, I managed to keep my weekends free for friendship and for birding. A Saturday morning in Bristol with mates usually involved wandering around the shops and trying on clothes, followed by a movie and a sleepover. The next morning, I would race home, grab my binoculars, and either jump in the car with Mum and Dad to twitch a , rare bird or go birding locally—alone. Why not combine birdwatching with my social life? I would have rather melted into a puddle than ask any of my friends to join me. I was a normal kid in that I wanted to be like other normal kids who didn't necessarily have any weird hobbies they scurried away to do on the weekends. I enjoyed my friends and I enjoyed my activism—so long as these two worlds never clashed! I was also performing to an adult audience, a different demographic from my peers. Among my friends, no one even knew I was Green Ambassador for Bristol—or if they did, they never let on.

I was very comfortable with my growing activism, but there was no way I would have given a talk at school about climate change or birding: the very idea made me feel cold. But the joke was on me, really. Everyone knew I was a birder and was aware of my column. So, of course, I knew they knew, and they knew I knew they knew. By now, the unwritten rule was that we never talked about campaigning or my birding life. And who decided this rule? Me, of course.

Australia, we agreed, had been a good trip, but the endless hours of driving and the constant anxiety around accommodation had been tough on Dad. Mum's moods were still erratic, which was unusual because the moods of most bipolar sufferers usually stabilize once they begin taking lithium. Dad needed to be sure Mum was on the right track before we embarked on our big trip to Uganda, Rwanda, and Kenya in the summer of 2015.

Together they visited Mum's psychiatrist, and Dad recounted the rows in the various camper vans, Mum's insistence on him driving crazy distances, and the general lack of organization. Mum was given a further diagnosis of *severe* bipolar disorder because the lithium had not helped "enough." She was put on a drug called phenelzine, an antidepressant, which has to be kept in the fridge. Weirdly, people prescribed phenelzine are advised not to eat cheese, given the risk of a dangerous rise in blood pressure potentially causing a stroke.

In Uganda, we hoped to see a Shoebill, one of the most desirable species for world birders—because it is not only very rare but also weird-looking, a scary bird. The Shoebill is also extremely cool, like a stork with a huge shoe-shaped bill. But what really rang our bell was the fact that it is monotypic—the only bird in its family—and is therefore unique and looks like nothing else. Birders are down to see any bird, but weird birds even more so.

My mum's parents are from Sylhet District in Bangladesh. Sitting in the Ganges Delta, between India and Myanmar, Bangladesh is perfect birdwatching country and also part of my heritage.

Ayesha spent her early childhood within the bosom of my extended Bangladeshi family. Mum was on her own with a baby; she needed their help, and they gave it willingly. Ayesha was raised by not only Mum but an army of aunts and uncles and my grandparents too. Our *nanabhai* loved her so much he would often bring home her favorite dish from his restaurant—tandoori chicken. In many ways, her childhood was entirely different from mine.

While my grandparents had eventually welcomed Dad into the family—he wasn't the first White boy whom Mum had married—it was still a bit of a shock when she introduced him to them as her fiancé. But so long as he was willing to look after Mum *and* Ayesha, they were happy enough. Mum and my sister had lived with my grandparents for a period, but now Dad was around, and my own experience of my extended

Bangladeshi family wasn't nearly so intense as Ayesha's. I felt, like so many children from immigrant families, as though I was growing up between two cultures, never fully belonging to one or the other.

But my identity crisis was resolved each time we went to Bangladesh. There, I am treated as one of their own. There is a Bangla word, *bilati*, that is used to refer to relatives and friends who live in Britain. There is lots of love and affection in this word and within these long-distance relationships. My Bangladeshi relatives understand that young people growing up abroad will have assimilated a new culture, and they make allowances for the subtle and not-so-subtle ways in which we're different. The color of Dad's skin has never mattered much to them, and I am always treated as just another family member who happens to have been born and raised in the UK.

Everyone has expectations about dual-heritage children and how they resolve this feeling of identity. But essentially, the confusion I felt had more to do with how people reacted to *me*; I was Asian to my White friends, and White to the Asians. With every visit to Bangladesh, I made a conscious effort to reconnect with my family, and I began to feel more comfortable in my skin by actively choosing to explore my roots.

And because Bangladesh is so important to me, so are its birds.

Sonadia Island in Bangladesh is one of the locations the Spoon-billed Sandpiper chooses to migrate to during the winter, and in February 2015, Mum and I headed over to join a task force to make a count of the birds on the southern coast. I had decided to use my growing platform—my blog had hit almost a million views—to campaign for the vulnerable bird. Given I am also Bangla, I leveraged my platform and ethnicity to gain wider attention.

The last time I was in Bangladesh it had been, as usual, a whirlwind of aunts, uncles, and cousins that involved driving long distances between towns and villages to make sure we had connected with every one of what felt like at least a thousand relatives. This trip was going to be very different.

We were picked up at the airport by Mum's uncle Josim. He hadn't seen her since 2011, when he visited her in hospital in Bristol. They laughed about how she had been as high as a kite the last time he saw Mum in Bangladesh. High on mania, that is. Mum was much better, but I couldn't help recalling that visit, picturing Dad waving us off as we boarded a plane to go home, only to have to follow us a few days later because Mum was so ill. I pushed this thought away—Mum was good, and Dad wouldn't have let us come if he was in any doubt about her health.

With hindsight, Mum has since questioned whether this trip was a little reckless. Not because she was ill but because we arrived in Bangladesh at a time of deep political unrest. They were still clearing the scorch marks off the tarmac outside the airport when we arrived in Dhaka—a minibus full of people had been set on fire the day before by violent protesters. The recent election had been disputed, and the country was in turmoil. Roads were blocked, and protests filled the streets. It was a tense time to travel, but we reasoned that we would be safe out in the mudflats of Sonadia Island. It hadn't been an easy decision to reach, but both Mum and I were invested in the project and promised Dad we'd confine ourselves to our hotel room until we could catch a flight out of the city.

From Dhaka we flew to Cox's Bazar on the southern coast, and from there to the mudflats and wetlands of Sonadia Island to begin the survey. Cox's Bazar, usually buzzing with tourists, was deserted; the town felt very empty and a little eerie.

Just like the geese that arrive in the UK during the winter from Greenland or Scandinavia, the Spoon-billed Sandpiper leaves the Arctic tundra in Russia, migrating to countries such as Bangladesh, where the food is plentiful during the bleaker months. With the destruction of the mudflats in Russia and China, the conservation of the bird became more difficult, which is why international cooperation was so crucial if the habitat of this sandpiper was to be saved. The project felt urgent, and for the first time, I was personally involved, however tiny my contribution, in helping a species thrive. While I had thrown myself into campaigning for the Spoon-billed Sandpiper project on social media, there was something very moving about physically interacting with these precious birds. It was peaceful. The reality behind the public campaigning was brought home to me as I stood knee-deep in mud. Everything was reduced to the simple action of counting tiny birds.

We walked the mudflats all day, just as Dad had done in 2011, until we had counted nineteen birds. I also saw a very rare wader and a new bird for my list: the Nordmann's Greenshank. Easily twice the size of the Spoon-billed Sandpiper, its call is a single melodious note. Dad had seen one in exactly the same spot four years earlier. Could it be the same bird? Quite possibly.

That first morning, I had felt anxious we wouldn't see any Spoon-billed Sandpipers, or maybe see just a couple, but while it had been a long, hot day, it was also hugely successful. High fives all around! There were only two hundred of these birds in the world, and, amazingly, we had counted a tenth of them in just a single day. There was some sadness in this excitement, however: just *two hundred* of these birds in the entire world, such a pitiful number. But at least the population seemed to be stabilizing after a dramatic decline.

There are many reasons why the Spoon-billed Sandpiper is so endangered; some of them are to do with Bangladeshi wader trapping—people are poor and the waders are plentiful. Unfortunately, the tiny bird tends to get swept up with the other, more abundant species. The trappers aren't intentionally going after the Spoon-billed Sandpiper, but nor are they going out of their way to avoid them. When trapping is your livelihood, even the tiny bit of flesh on a small bird is valuable.

No one *wants* to be a bird trapper—it's not a lucrative or respected job—and part of the project was to encourage different lines of work. Out on the flats, I met a local man who had been a trapper for many years, unable to imagine any other way to make his living. The project sponsored him to use his other skills: As a child, he had helped his mother take in sewing from the local community. Soon he was set up in his own shop and went on to become a successful tailor. Another huge concern is that all along the East Asian coast, there has been a loss of habitat from the industrialization of the wetlands.

Why did I feel I had to go all the way to Bangladesh to do the survey work when there were already others out there working on the project? The single useful thing I could bring to the project was media awareness: given that my blog had a profile now, and even though I was a child, I had strong opinions about conservation, and I wanted to use this profile to help to save the Spoon-billed Sandpiper, to talk about the threats to its survival and the reasons why its conservation was so crucial. I gave interviews in the Bangladeshi broadsheets and appeared on a popular news channel. During this period, my blog was viewed by tens of thousands of people.

Many conservation projects begin with selecting a single species, usually an iconic bird or animal, one that is unique or beloved, a creature people can really get behind. You focus your efforts on a flagship animal, and as a consequence of the campaign, other birds, animals, and insects sharing its habitat are also saved.

Since the Spoon-billed Sandpiper project began, the number of fledglings that survive each year has increased by 20 percent, much of it because of headstarting—the process where birds are hatched in captivity and released only when they are bigger and stronger. But the real takeaway is that as soon as you have a big project focused on saving one species—and this one was huge, with many different countries coming together, including Russia, China, South Korea, Thailand,

Bangladesh, and the UK—then the results can be overwhelmingly positive. The obvious issue is that such projects just aren't feasible for the vast number of troubled species.

The Spoon-billed Sandpiper is slowly recovering, but for how long? As global warming impacts the land—and however bad we think it is, the reality is much worse—the international effort for the Spoon-billed Sandpiper becomes all the more poignant when you imagine that if all our glaciers melted, the global sea level would rise by approximately seventy meters, flooding every coastal city on the planet.

This visit also coincided with my growing awareness of race and diversity, especially in the area of conservation and nature. In the UK, the overwhelming majority of people who work in the sector are White: birders, scientists, and researchers. My extended family never understood my desire to spend time in nature, and I had accepted it as a cultural difference, but this trip to Bangladesh opened my eyes: I met passionate young birdwatchers, campaigners, scientists, *and* researchers—all of them Bangladeshi. Why was this? Why were there thousands of Bangla birdwatchers here, yet in the UK, I knew only three—my mother, my sister, and me? Why were there so many birdwatchers in Bangladesh but none in the diaspora?

"Mya, there are hardly any *women*, never mind Bangladeshis." Dad's take on the ethnic disparity in the birdwatching community was enlightening. My family was *already* a pretty unique twitching combination, an anomaly almost—husband, wife, and daughters—even before considering ethnicity.

It was clear there was some way to go.

In Dhaka, Mum, with the help of the Bangladesh Bird Club, had organized an event during which I would raise awareness around the Sundarbans oil spill and the Spoon-billed Sandpiper.

While speaking to a room full of nature lovers, conservationists, and members of the media, I thought that if there are lots of young people passionate about birds, nature, and conservation here in Bangladesh, there could be no reason for a lack of campaigners from a similar background in the UK other than exclusion, and that this "othering" of visible minority ethnic (VME) people is causing their disconnection from nature. Of course, they might have no desire to go birdwatching, hiking, or any of the other thousand things you can do outdoors, but they should all be given the choice.

I have chosen to use the term *visible minority ethnic* in my work and my activism because it is a very useful indicator, especially in the nature sector. Simply put, VME refers to minority ethnic people who describe themselves as non-White. BAME (Black, Asian, minority ethnic) is a more common term but not terribly useful in the sector because, when it comes to going out into nature, the barriers that exist don't apply to many minority ethnic people, as they are White. The barriers exist because of the color of your skin, because you look different from everyone else.

If the figures for BAME workers in the nature sector are low, drilling down we discover that the numbers of *non-White* employers are catastrophic, just 0.6 percent. The term *BAME* can be used to actually mask just how bad the situation is for VME people.

An idea began to brew: one way to tackle the imbalance would be to address it outright, to act on it, and to force change. I had my platform, a place to discuss the issues I was interested in, but blogging can achieve only so much—the echo chamber of the internet was loud even in 2015. I needed action, right now.

I had recently become aware that in America, there were kids' summer camps for everything, from karate to Christianity to birdwatching.

"I'd like to go on a birdwatching summer camp," I told Mum.

"In America?" she said.

"Not in America! Here."

"No such thing, but you could start one," she suggested.

This was Mum all over. If I wanted something, I had to make it happen, just as she had done her entire life. She became a solicitor when there were few Bangladeshi women even attending university. She had worked hard to ensure that her law firm hired more people from VME backgrounds. And when it came to birdwatching, Mum wasn't ever going to be thwarted by time pressures, reluctant guides, or impending nightfall.

I wanted to hang out with kids my age, who looked like me, who cared about what I cared about. Bristol is, after all, one of England's more diverse cities, with high immigrant populations from South Asia, Africa, and the Caribbean. What was to stop me from forming my own summer camp? The audience was already there; I just had to ask them to come outside.

A successful word-of-mouth campaign, coupled with some social media advertising, attracted fifteen teenagers for our first summer camp at Avalon Marshes, near Glastonbury. But…they were all White boys from a similar rural, middle-class background, and I immediately regretted the direction of our pitch. Instead of a *birdwatching* camp, I should have emphasized the outdoors and camping, something that might appeal to a wider group of inner-city kids. An invitation clearly wasn't enough. If this was to work, I had to proactively engage VME kids. If I ever had a light-bulb moment, this was it. I had no idea why they weren't coming, but I understood that I couldn't just parachute teenagers into an environment they had little awareness of and somehow force them to unlearn the message that the countryside is not for them.

Mum took the reins, marshaling Bristolian volunteers to run the activities while reaching out to parents of VME children in an effort to persuade them to send their teens our way. In June 2015, we ran the first weekend Camp Avalon for twenty teenagers.

It's true to say it was stressful and not entirely successful. On the first evening, while I was pegging down a tent, I overheard a couple of teenagers who were inside.

"Your mum make you come?" asked one.

"Yeah," said his friend. "You?"

"Same."

I was mortified.

The first camp featured a whirlwind of activities, some more successful than others. The pollen count was high, and the hay fever sufferers were sneezing. Some of the kids were freaked out by the mothing session, and there were lots of complaints about the long mammal-tracking walk. No one at all enjoyed being woken up at six o'clock on the first morning for a hike on the Somerset Levels.

While I had managed to fill the places, it was clear that the five Black and Asian teenage boys who had come along had attended only because their mums had made them.

What the hell have I done?

Clearly I had misjudged the challenge, which was how to make the countryside exciting to urban youth.

"Who the hell gets up at dawn to go for a walk?" they asked as we trudged along.

"You can't be tired," I said. "You all play football, for God's sake." Blank stares. No amount of banging on about the benefits of fresh air and the wilderness from Mum or Dad did anything to change their minds.

In the end, it was one of our local nature-loving volunteers who turned the camp around. I watched the teens sit up straighter as he began to talk about the Peregrine Falcon. Here was this cool dude in his twenties comparing the falcon to Formula One race cars: both powerful machines, easily accelerating to speeds of up to 370 kph. The falcon, storming the skies like a firework, thinks nothing of swooping down on unsuspecting pigeons and crows come dinnertime. It is a fighter plane, a race car, a rocket—and luckily for my camp, the gateway into the wonders of birdlife. The speed, grace, and power of the Peregrine Falcon marked a turning point for future camps: making nature relevant to the lives of the campers.

After we had packed up and said goodbye, I felt exhilarated. In the end, the camp was a success; everyone had connected with and enjoyed nature to some degree. I was more determined than ever that there should be a place in nature for all ethnicities. But I realized that if change was to happen on a national level, wider action from conservation communities would be essential. I had also learned that our volunteers were crucial to a successful session. They had to connect with the kids by making nature relevant. That is the first step in breaking down barriers.

Later in the year, I began writing to major nature organizations, such as the RSPB, the Wildlife Trusts, Wildfowl & Wetlands Trust, and the British Trust for Ornithology, hoping to start a conversation about their inclusion policies and what they were doing to encourage diversity engagement in nature. I received responses from every single one. They were very interested in Camp Avalon, but they themselves were doing nothing, and would I go in and talk to them about how they might improve their communication strategies?

It was obvious they needed to talk to someone, but I was just a thirteen-year-old schoolgirl with a dream. I would have to dig deep to come up with the goods, but I could feel change coming.

8

Don't Mention the Chimps

SHOEBILL

The Shoebill is a resident of the freshwater papyrus swamps of east-central Africa. With an estimated population of five thousand to eight thousand individuals, it is classified as vulnerable. The main threats are habitat destruction or degradation, hunting, disturbance, and illegal capture. It is a huge bird, standing up to 1.5 meters tall with its enormous, characteristically shoe-shaped bill. It harbors a gruesome secret: when nesting, most pairs produce two chicks hatched five days apart. This means the firstborn chick is significantly larger than its sibling. While the parent birds are away from the nest, the larger chick will physically attack the smaller, drawing blood and driving it to the edge of the nest. The parent bird appears to support this activity, upon its return shading the larger chick with its wings and providing it with water from its bill while the smaller chick is left out in the heat to die. The sole purpose of having the second chick is as a backup in case the first chick doesn't survive.

In a taxi streaming through the nighttime city streets of Kampala—mopeds screeching all around, the bright lights of restaurants and bars glowing neon in the dark—I was trying to get a feel for Uganda, for the heat, the humidity, the exhilarating smell of another country. But given that we weren't staying in the city, I settled back in my seat and shut my eyes.

East Africa had been selected by Mum and Dad for not only our target birds but its "latitudinal diversity" as well. Areas between the tropics have much higher species diversity to start with, and if you add in vast areas of pristine habitat such as savanna, wetlands, rainforests, and the odd mountain range, the sheer number of bird species is a magnet to birdwatchers.

The Shoebill topped all our lists that summer and is seen as a prize for not only serious birdwatchers but wildlife fans in general; everyone is fascinated by its size and *strangeness*. Mabamba Swamp is the most accessible and reliable place to spot the Shoebill in Uganda, and probably the whole of Africa, and on our first morning, that's where we were headed: the Mabamba wetlands on the shores of Lake Victoria.

I had been desperate to get away that summer. As I approached the end of year eight at school, I took stock: campaigning for endangered species and Camp Avalon aside, while I'd made new friends at secondary school, it was obvious that many of my peers believed birdwatching as a hobby was a bit weird. The social confidence that seemed so effortless to everyone else was something I struggled with. That's how I saw it, anyway. With hindsight, I should have faked a bit more bravado, as it's clear to me now that's what everyone else was doing.

Ayesha's advice to me when I started secondary school was to never mention birding if I wanted to avoid being teased, but unlike her, I couldn't hide it entirely, because of my social media presence. Local press and occasionally the national press had begun to approach me for interviews and features about my birding experiences. I never seemed to lack self-confidence when talking about birds or the environment. Of course, some schoolmates were also aware of these media events. I had started to dread IT lessons because invariably one or the other of my classmates would look up my blog or a recent interview and begin to read aloud.

"'It is amazing how many species live in one garden...,'" someone would begin.

"'Just sit in your garden and get spotting,'" another would add.

Our IT teacher rarely intervened, maybe because it wasn't bullying so much as aggressive teasing. But I would always react, and it didn't take me long to figure out that this is exactly what my tormentors wanted. I'd usually lean over and delete the tab with the article, despite the cardinal rule in school never to respond to your persecutors. I broke it again and again. They'd pull the article up once more, and I would sit there squirming.

The takeaway here isn't that my peers were really nasty but that I was desperately uncomfortable in the school spotlight. On the day the teasing stopped, it wasn't because they had matured and moved on but because I no longer cared what they thought. I got over myself. But this was some way off.

If school was stressful, local birdwatching was more so. Somerset is great for birding. Chew Valley Lake itself is famous for the rare birds that have visited over the years, such as the Booted Eagle—this happened before I was born—and various rare ducks, herons, waders, and other waterbirds. Situated along the main road around the lake are two of the main birding hot spots in Somerset, and this is where I often hung out on the weekends, usually in a bit of a panic in case a friend driving past should spot me. Whenever that happened, I would look away, but they would also look away, both of us equally appalled to find me wandering all over the grass verges, binoculars aimed into the trees. Birding had become embarrassing. How could something that gave me so much pleasure become a source of tension?

Paradoxically, for all my edgy self-consciousness, I was also unwilling to give up even a moment of it. Away from the roadsides and in the wilds of the countryside, I could forget about school, stop worrying about Mum, and lose myself in the very simple game of waiting, watching, and spotting. Birding was my respite, and it was becoming increasingly obvious I was going to have to figure out a way to integrate it into the whole of my life.

But right now, I didn't want to think about any of that. I was climbing into a boat on the Mabamba Swamp; school and the Chew Valley had evaporated. In a few days, Digby would be joining us for three weeks of intense birdwatching in Uganda. Would he be wearing his green vest? I wondered.

The extensive papyrus swamp, with its labyrinthine channels and lagoons, had been classified as an Important Bird Area, and it was home to several pairs of Shoebills, Uganda's most famous avian resident. This charismatic monotypic species is certainly among the most sought-after birds in Africa, and we were going to make an extra special effort to find it by paddling through the channels of thick marshland by motorized canoe.

The favorite food of the Shoebill is lungfish, plentiful in the swamp. However, the lungfish is equally valuable to the local fishermen, who had long held the superstition that seeing a Shoebill would result in a poor catch that day. They hunted the Shoebills and killed them, driving them to near extinction in the wetland. After that, Mabamba Swamp was designated a Ramsar site (an international initiative to preserve vulnerable wetlands) in order to protect the Shoebill. Today, many of the local fishermen rent out their boats to birdwatchers, and some have even retrained as bird guides. In this way, the fishermen protect the Shoebill rather than view it as a bad omen.

After some minutes in the canoe, we saw the papyrus reeds open up into the big grassy swamps. This is where we would find our bird, apparently. Given their out-of-proportion features and massive dirty-yellow bill, resembling a battered old shoe, I didn't think they'd be too hard to spot.

Apart from the frogs, crickets, and occasional birdsong, the marshes were eerily quiet. But it was hot—the blazing African sun laughed at the stupid birdwatchers, sweating in their boats.

When the Shoebill felt the heat, it had a solution unpalatable to the rest of us. To cool itself down, it practices something called urohidrosis, where it defecates on its own legs and the subsequent evaporation cools the bird down. I decided I would need to be quite a bit warmer to do the same.

And then, at exactly midday, we received a call from another boatman, giving fuzzy directions over the radio, urging us to a particular spot among the reeds. Our boatman knew the maze of reeds like he knew the corridors of his own house, and soon we approached the spot to find other boats full of excited twitchers. This was an unusual experience for us—usually we're alone, standing around, waiting for something to happen. It felt good to share these moments just before a big twitch with other birders.

We motored forward, following the sound of clicking cameras; everyone had their lenses trained on a very strange creature one hundred meters away. A mainly gray bird, this storklike dinosaur was posing for pictures. For a moment I was startled: the bird wasn't moving or even blinking. Despite the fact that its enormous shoe-shaped bill remained clamped shut, it looked like it wanted to eat us. With a wingspan of over two meters and a height hovering around the one-and-a-half-meter mark, it looked powerful and dangerous. For a long time, everyone stared at the Shoebill and the Shoebill stared right back. Eventually, it opened its bill, cracking a threatening smile before dipping down to the water. It hadn't been interested in us at all; it was waiting for the juicy frog, which was now clamped in its mouth. Every single person in every canoe gave a gasp as the Shoebill spread its enormous wings and took to the air.

Mum high-fived Dad. "This is more like it!" she yelled. She was beaming, squinting in the sunlight, and she looked absolutely delighted with life, in a boat, in the middle of a swamp.

Mum's head is generally a very busy place, switching off for a few moments only when her target bird appears. The Shoebill had stayed with us for an hour, and Mum watched it all that time, clearly in a state of wonder. In moments like these, her looping thoughts can go as still as the Shoebill as her mind finds its focus.

Was there an element of flight *away* from our real lives on our trips? An escape? It's a question we asked ourselves, and time and again the answer was no. While Mum usually thrived on our holidays, her illness came with us, ever-present. Our desire to see the birds was what drove us, and we had proved that being in nature helped both Mum *and* Dad. While traveling had never provided a solution to Mum's bipolar disorder or Dad's coping mechanisms, it had hugely improved our lives together. Today, we have no expectation that traveling or anything else is going to make Mum's mental health situation go away, but we are better as a *family* when we do it. Most important, our trips shore up the rest of our lives; the birds tend to linger long after we have walked out of the rainforest, left the savanna, or climbed off a boat.

Now thirteen years old, I understood that there was to be no single magic pill for Mum's bipolar disorder, despite the diagnosis. While her meds were under constant review—her moods continued to fluctuate—we were all also adjusting to the fact that her brand of bipolar disorder was more about long-term management to prevent a full relapse and less about a pharmaceutical one-shot solution. I was getting used to the idea that she would always experience dramatic mood swings. It was upsetting, and I still got angry with her, but I was also getting older, becoming a little more self-reflective.

Mum loved being away, and it was obvious it did help her. The more remote, the rarer the birds, the better. But while traveling suited Mum down to the ground, one of her meds had to be kept chilled. This is of course easy at home with a fridge, but not in hotter climates when you're constantly on the move. For Africa, Dad planned a system of cold flasks in which to store them. Inevitably, we wouldn't always be successful, which meant the pills would be less effective, and then anything could happen.

Digby joined us in Uganda, having spent the previous day at Mabamba Swamp catching up with the Shoebill. We were even stephen! And, yes, he had on his many-pocketed gilet.

Kidepo Valley National Park nestled in the wild frontier region of the extreme northeast of Uganda, bordering South Sudan. The open-country grassland, home to lions, elephants, and buffalo, was also the habitat of the Karamoja Apalis, our target bird.

The park is likewise home to the decaying remains of Idi Amin's infamous safari lodge. The dictator, commonly known as the Butcher of Uganda, is regarded as one of the most brutal despots in world history, thinking nothing of expelling an entire ethnicity from Uganda and slaughtering more than three hundred thousand of its citizens. We had Asian Ugandan friends who had had to flee the country back in the 1970s, and here was the house of the man who had sent them away. It was a strange and unsettling experience, at odds with our mission to locate rare birds in the wilderness.

In the Land Cruiser, with guide and driver, we turned our attention to the Karamoja Apalis. By the time we arrived at the site, it was the middle of the day: boiling hot and less-than-ideal conditions for searching for the small gray bird in its bushy savanna home.

When we're stumped like this, we'd usually climb out of our vehicle and head off the track to look for other birds, but this was lion country. We did get out of the car, but stayed mostly within touching distance. At least, that was the plan; Mum was moving farther into the bush.

"Helena, come back," Dad pleaded, but she wasn't listening—she was desperate to see the apalis.

"Stop where you are!" bellowed Digby. Mum froze. "Do you think we want to sacrifice three days of birding just because you've gone and got yourself eaten by a lion in Africa?" He figured it would take Dad at least half a week to do the paperwork for Mum's untimely death by big cat.

Mum had to laugh—we were all laughing. She came back to the car and we decided to move on, birding our way toward the South Sudan border.

As we approached the border, we passed a turning off the main track, which led to a small Ugandan military encampment. There was no one there, and so we plowed on. A short distance later, we arrived at a dried-up streambed crossing the track we were traveling along. This was a border of sorts, and on the other side was South Sudan—no barriers, no border guards, not even a sign saying "welcome," and we had no visa to cross.

"Let's go over," said Digby. An innocent enough suggestion, but South Sudan, at the time, was engaged in a violent civil war.

This fact didn't deter anyone but our guide and driver. We were suddenly very keen to add South Sudan to our country list (yes, we have those as well), and Digby and my parents egged on our driver. Dad was particularly keen to visit a country that had only gained independence in 2011, making it the most recently recognized sovereign state in the world and us most likely the first world birders to visit it. (Sometimes we just can't resist a bit of exclusive one-upmanship.) But our driver pointed out that this was "bandit country." I felt a thrill of fear, but Mum and Digby were by now raring to go.

We were behaving like stupid gung ho tourists, obsessive birders who would do anything to add a new bird to our world list, any bird as long as it was seen in South Sudan. We didn't listen to either our driver or our guide, who were the only ones who really understood the risks of crossing the border. If anything happened to us, they would be responsible.

After twenty minutes and no new bird, driver and guide finally cracked and insisted on turning the car around. We moved back up the track toward the dry stream when a thrush-sized bird flew out of the line of trees that hugged the edge of the stream, crucially on the South Sudan side. We had our bird, a Slate-colored Boubou. If our driver was anxious, he almost erupted when we all exited the car to try to take photos.

"You get in now, or I leave you here!" he insisted. Dutifully, and a little shamefaced, we climbed back inside. The boubou returned to the trees, leading us back to the border, and our driver breathed a sigh of relief, probably thankful the bird was taking the stupid tourists back to safety.

This small gray bird was at liberty to fly wherever there was food, with no need of a visa or an awareness of civil war. Never had the freedom of birds felt so real to me. All it wanted to do was survive, yet on the ground, we were willfully neglecting the planet to such a degree that very soon its home might become so degraded it would begin to struggle.

Nevertheless, the nerd in me found this sighting beyond exciting, for the sad fact that we were seeing it illegally.

Pittas are difficult birds to see, despite coming in shades of emerald green, violet blue, and bright yellow, some even adorned with fire engine–red patches. They are what's known in the business as "skulking" birds, preferring to hang out in dark spots of undergrowth, keeping very still, as opposed to showing off their glorious feathers in the trees and in the air. They require patience, which, as birders, we have plenty of.

Kibale Forest National Park, in Uganda's Western Region, is noted for its stunning habitats of wet and dry rainforest and savanna. It is famous for its chimpanzee-trekking safaris. Crucially, the park is also fantastic Green-breasted Pitta territory.

Pittas are best seen at the crack of dawn, and so our expedition into the forest in search of our target species began very early. We arrived at the park rangers' offices at four thirty in the morning, raring to go, only to find the guides arguing among themselves about who would take out which of the groups.

"We could still be sleeping," sighed Digby.

"We could have had a go ourselves," insisted Mum.

Finally, we were allocated a guide, and we followed him into the thick, dark woodland to begin searching for a small bird. The forest was a somber place under its heavy canopy. Pittas tend to hop around on the ground rather than lurk in the trees, and we were lucky that there was only a scattering of bushes, which meant fewer places for them to hide. The walking was easy but frustrating, since the gloom made it so hard to see *anything*. As a small child, I wasn't a great fan of the waiting around that went with birdwatching, but it's true that patience comes with age, and while I was much better walking and waiting now, not being able to see one inch of blue sky was starting to make me feel claustrophobic.

An hour later, there was still no sign of the Green-breasted Pitta, and our guide decided we should make our way toward another part of the forest, this time by car. We climbed into his 4x4 and set off down a very bumpy track. All the while, it was getting later and lighter, and we hadn't even caught the scent of a pitta. We didn't fare any better in the second forest either. I decided today would be a washout. Standing around in the dark, insects flying directly into my mouth and ears, I was ready to get out and into the sunshine.

Our guide's walkie-talkie buzzed and he began to chat excitedly in the local dialect. As he talked, I heard a rustle in the branches overhead and my heart leaped—maybe another wonderful bird would come to our rescue and make the whole morning worthwhile. But it wasn't a bird. I was looking at chimps, lots of chimps, and they were staring right back at me. If I had imagined chimps in the wild at all, this wasn't what I expected. Shouldn't they be screeching or swinging wildly from tree to tree? These guys were mellow, offering us no more than a grin.

"We have to go back to the first forest," our guide announced, pocketing his walkie-talkie. He explained that the chimp guide had just radioed him to say a pitta had been spotted and we should return immediately. At that very moment, three chimp-seeking vehicles appeared on the track. We were all about to flag them down and point to the chimps hovering above our heads when our guide gave a quick shake of his head.

"Don't mention the chimps," he whispered. "You didn't pay to see the chimps, so I'm not allowed to show them to you."

"What?" I said.

"I think he'll get in trouble if the other rangers find out we saw chimps," Dad elaborated. "He's been paid to show us the pitta, not the chimps, Mya. We'd be getting something for nothing, and that isn't allowed. His boss wouldn't like it."

The chimp group had paid for their chimp permit and were therefore allowed to be shown the chimps—just not by our guide. I felt sorry for them as I watched their 4x4s disappear down the track in search of the elusive ape, when all they really needed to do was look up.

"But their chimp guide was allowed to tell us about the pitta," I pointed out, back in the car.

"I think we let this one go, Mya," said Digby wisely. "We're losing focus." Everyone was losing focus, in and out of cars and forests, illegal chimps and oblivious chimp hunters. I let it go.

Now, back in forest number one, we were at last making some progress. Dad had picked up the distant call of the pitta and led the way. We scanned the open patches of the woods ahead with our binoculars, but the bird wasn't anywhere on the ground, where it was supposed to be lurking. Our guide was now striding ahead, farther into the forest, keen to reach a familiar haunt of the Green-breasted Pitta before it got too late. But we were all feeling tired—it was so hot in here, and it felt as if we were walking around in sweaty circles. Our guide eventually stopped walking. I guessed we had reached the site.

"Wait here," he whispered, heading off into the trees in a loop around the clearing, periodically playing the pitta's call.

"I think we should explore too," said Digby, never one to stand still. He was clearly frustrated and wanted to go off and follow the song, not the guide.

"I agree," said Dad, and off they went, deeper into the forest. "If we find anything, we'll call you."

"If you don't flush the bird before we get a chance," I snapped as they disappeared into the trees.

Mum and I were left alone, feeling very small in the middle of a vast forest. What wild creatures lay in wait for us? And then we heard it, the distinctive call of our target bird. It was close, really close. This was no recording; it was the real deal. As Mum and I headed toward the birdsong, our guide appeared behind us, and Dad and Digby materialized from the jungle ahead. We were all converging on the same spot. This had to be it.

The trees parted, revealing another small clearing, where, perched ten meters up in a tree, sat the Green-breasted Pitta, waiting patiently for us to arrive. Digby switched on his camcorder and caught the pitta's melodious hello as he puffed out his glossy shades-of-green chest. The trees weren't so dense here, and a little sunlight poured in, catching his fire engine–red belly, the delicate pale blue spots on his deep emerald wings. Now that we had arrived, the tiny bird was ready to begin his mating dance. A lone branch bisected the clearing, and the bird hopped down, deciding this was the perfect platform from which to launch himself on and off, all the while calling and flaring and whirring his feathers, declaring, *This land is my land.*

It took a long time for birding to become a mindful exercise for me. When I was a kid, it was fun and exciting. I never thought about being "present" or trying to stay focused. There were always more birds to see, and I was eager to be off to the next hot spot. I would fidget, chatter, and generally want to get going soon after we had seen what we'd come for. I was thirteen years old now, better able to appreciate standing still, shutting off my mind to the outside world in the presence of a spectacular bird. I had begun to enjoy the anticipation of waiting for our target to appear, those moments when you're not aware you're even breathing.

If I was to pick a top mindful moment, it would be that encounter with the Green-breasted Pitta. The insistent voice in my brain that can't help naming everything it sees and hears went silent.

Deep in a Ugandan rainforest, watching a small bird sing and dance its heart out to an audience of five humans, my own heart felt like it was exploding. Everything else in my life receded in a second. This was better than I could ever have hoped for—a few magical moments watching a display of extraordinary beauty. Tears pricked my eyes: there wasn't anywhere else on earth I'd rather have been than here in the clearing watching this little bird call for a mate.

Birds are wild animals. They are mercurial creatures of raw biological power. The robin in the hedgerow is as wild as the Green-breasted Pitta dancing for us in the clearing. My garden, this forest, and the wetlands of Sonadia Island in Bangladesh are life-sustaining habitats for birds, and as long as they thrive, the birds will come. It feels like an obvious thing to point out, but it's worth repeating: nature organizations of course know a thing or two about birds, but their *expertise* is landscape and conservation. If you get the landscape right—if the habitat is welcoming—the birds will come. There is no other way to get them there. To engage with birds is to engage with landscape and with nature; for me, they walk hand in hand. And those powerful moments in the rainforest, watching a colorful bird instinctively attract a mate and make more Green-breasted Pittas, helped me see, very clearly, that environmental conservation is vital to their survival.

Digby's three weeks were up, and we waved him goodbye. He was leaving to join his own family on holiday in Europe. Again, I was sad to see him and his gilet go. His company had been good for us, and he hadn't needed to raise the issue of the Craig Family Harmony Index once!

We were on the Ugandan side, waiting to cross the border into Rwanda. The queue was long and busy, car horns blaring, border guards shouting for patience while they checked papers in the dusty heat. Motorcycles sped up and down the lines, their riders selling cold sodas and hot snacks. As soon as we were waved through, our surroundings changed. Overburdened motorcycles are ubiquitous all over Africa, bearing entire families, usually helmetless, some even balancing pigs on the handlebars, but in Rwanda it was a different story. There were few cars, never more than two people on a moped, and everyone wore a helmet.

The Rwandan genocide of 1994, led by the majority Hutu population, was a brutal attack on Tutsi civilians, resulting in the slaughter of over eight hundred thousand people. But in the middle of August 2015, it was hard to imagine such a devastating campaign of violence had ever taken place. Here was a nation of people collectively suffering from post-traumatic stress disorder. But at the same time, this thriving country was spotlessly clean. I was fascinated to learn that once a month, at least one person from every household has to take some responsibility for clearing or maintenance work outside the home, whether it's to trim bushes, fix fencing, or sweep the streets. This custom is known as *umuganda*, "community service." Initially introduced to combat the ruinous effects of war on the country, it remains today, making Rwanda one of the cleanest countries in Africa. We passed convicts in the fields dressed in crisp, clean orange uniforms, better dressed than many of the people in the poorer areas of Uganda. Something felt a little strange, though, but I couldn't quite put my finger on it.

Nyungwe National Park, in southwestern Rwanda, borders Burundi to the south and the Democratic Republic of the Congo to the west. With almost one thousand square kilometers of rainforest, grassland, swamps, and bogs, it is considered to be the best preserved of the Albertine Rift forests. It is a fantastic stomping ground for its three hundred bird species and provided spectacular views of several species we had already managed to glimpse farther north, in the forests of Uganda.

But there was one target species that, realistically, we were not going to see anywhere else in the world: the Red-collared Babbler. While it was not technically endemic to Rwanda, even Dad wasn't proposing crossing over the border into Burundi or the Democratic Republic of the Congo to look for a bird; this was our only chance. We trekked through the forest on multiple lengthy trails, our local guide taking us to the best and most reliable sites, but by the end of the second day, there was still no sign of the Red-collared Babbler. Tired and thirsty after another long hike, we returned to our vehicle parked beside the road.

After a quick stop for a drink and some much-needed biscuits, we decided to give it one last go. There was a short trail parallel to and within earshot of the road. It wasn't the most promising bit of habitat, given the kilometers of pristine rainforest we had walked through. Literally five minutes along the trail, we encountered two Red-collared Babblers chattering away to each other while moving rapidly through the canopy. They were vivid streaks of color among the green leaves, their red collars clashing brilliantly with their black caps and pale bills and eyes—they were worth each of the thousands of steps we had taken to find them.

After a couple of days in the park, the strange feeling returned: something was definitely a little *off*. Rainforests are generally busy places, with birds singing, mammals scurrying, and the seesawing cries of insects. It wasn't silent, of course, but it wasn't *noisy* either. The savanna was even quieter. Where was the big game? The Rwandan genocide had happened years before I was born; a tragedy that had felt very distant suddenly came alive in the silence. During the period of unrest, desperate people—soldiers and civilians alike—had fled, starving, hunting whatever animal or bird they could find. Everything is linked in the wild, and hunting disrupts the whole ecosystem, not just the animals targeted. It might affect those that feed on carrion and bird migration and the pattern of hibernation for mammals. The sheer scale of the crisis depleted the wildlife to the extent that no big game or predators were left at all on the savanna.

I was seeing the scars of a long-ago war. The depleted savannas reflected the emptiness and loss left behind after the war. I had just been in Uganda, a country with thousands of animals on the African plains, identical habitats to those in Rwanda. It was as if these vast unpopulated regions were in mourning for the genocidal loss of human lives.

We had been mostly successful keeping Mum's meds cold as we traveled the hot grasslands and moist rainforests of East Africa, but not entirely. By the time we reached Kenya, one of Mum's pills had overheated and smelled so horrible that it had to be jettisoned. Just the day before, she had been reluctant to return to our lodge after a full day's birding, insisting there was more she could do and see. This was a warning sign, there was no doubt about it, but we were on the last leg of our trip: in twelve days we would head back to the UK. She just had to hold it together.

Kenya, with its colonial connections to the UK, stable government, relative ease of travel, and, most important, incredible diversity of habitats, big game, and bird species, has been a longtime favorite destination for British birders. Perhaps strangely, it started as a bit of an add-on to our main trip to Uganda. But now Mum, having realized we had just under two weeks to do the rest of our birding, was keen to maximize our time here.

The Swara plains are just thirty minutes from the bustling city of Nairobi. We drove out of town at five o'clock and into wide-open grassland. By this point, I was a handful of birds away from my four thousandth species, and keen to reach that milestone. Our Rwandan guide had desperately tried to find the six extra species I needed, but it was not to be. On my first morning in Kenya, I made it.

The sun rose as we walked among low bushes and scrubby grass, and there were birds everywhere. Part of me wished we hadn't started out so early on our first day—the trip was catching up with me, and by nine o'clock, it was already way too hot. But Dad is a machine—he was on the home stretch and hungry to see as many species as possible—and Mum wasn't going to be outbirded by Dad.

I just wondered which species was going to carry me over the line. I didn't care if it was rare, endemic, or especially pretty. I was feeling exhausted.

In the end, it wasn't a gorgeous bird that took me to my milestone but the Red-throated Tit, kicking around in the dust. With its furry head and neck of rusty-red feathers, this tit had just the right amount of style to earn its iconic position on my list. I had been building up to this moment for three years, and now it was here. But, as always, I didn't have time to linger over the landmark: there were so many more birds to see, and Dad was pushing us to keep moving. Later, when I got a minute to myself, I was as excited as the three other times I had reached a millenary milestone. Birding can be a macho game; it can be a bit competitive, and so can I. Despite my age, I now had the list to prove my worth as a world birder. There's nothing to stop you faking a list, but why would you?

Psychologically, the goalposts tend to shift once you've reached a big target. What feels like a monumental event as you approach it recedes fairly quickly once your goal has been hit, and then it's on to the next. With the Red-throated Tit I had moved up a level and was now among a group considered to be very serious birdwatchers. Each "thousand" milestone isn't just a bigger number; it represents a serious passage of time, telling the story of how far you've traveled and the huge effort you've put into birdwatching since your previous target.

Undoubtedly, the Regent Bowerbird was an altogether more striking landmark bird, but the Red-throated Tit was equally special to me, and not just because it was my four thousandth bird. It was also a symbol of the adventure I had begun as a very small child, a measure of how far I had come.

Dad's dedicated instruction served me well. During our travels, these concentrated periods of looking, waiting, and observing fed my desire to become a better birder. I learned the basics from him—how to locate birds in the trees, in the sky, at sea; how to recognize birdsong—and now, in the wild, I was putting those skills to use, honing them, enjoying how easily they came to me.

When I was younger, spending so much time with my parents on holiday had been fun, mainly because for much of my early childhood, they had been too busy with work to take extended holidays. But I was thirteen years old, and my adolescent need for some space was growing. When it was time to go home, I was ready.

So was Mum, even though she wasn't aware of feeling unwell. It was becoming obvious that a manic episode was right around the corner. She was impatient, quick to anger, and reluctant to go to bed. It wasn't her meds that needed adjusting; she just had to return to her prescribed dosages, which hadn't been possible owing to refrigeration issues while traveling.

Our trip wasn't over the moment we entered the cottage. In the months that followed, we shared our adventures with the family and relived our Ugandan highlights with Digby. (We still talk about Uganda with him to this day.) Mum poured her energy into writing up and posting our bird lists to the BUBO website. We had spent weeks and weeks outside in nature; it was regenerative, physically energizing. We had a good stock of positive energy in our banks for the challenges ahead, sustaining us far beyond the moment we landed back in the UK.

• • •

Africa had taken me out of my life for a while, but now I was back, and the familiar anxiety about my extracurricular work took hold once more.

Concern over my visibility as Birdgirl to my classmates had melted away in Africa. I was aware that, for the first time, my trip had been about not only exploring the wonderful world of birds but also escapism, pure and simple.

No one asked me what I did when I disappeared over the summer months, and I assumed this was because they knew I was doing something so deeply uncool it mustn't be mentioned.

But now I was determined to step out of the shadows, so I posted some pictures of the safari animals I had seen in Africa on my personal Instagram; it felt like a big step and a bold move, and most significant, there was no mention of birds! I was just a regular girl posting about her extraordinary holiday.

Only one friend responded to my post: she told me her mother had "loved it." It wasn't quite the reaction I wanted, but I learned something valuable. I now understood that it didn't matter where I went or how I spent my time when I was away—whether I was campaigning for climate change or organizing VME camps, no one was *that* interested anyway. Teenagers are just generally more into themselves than anyone else. Wasn't I the same?

9

Voyage to the Bottom of the World

KING PENGUIN

King Penguins are the second-largest penguin after the Emperor Penguin, mainly inhabiting subantarctic islands, including vast, dense colonies of several hundred thousand breeding pairs on South Georgia and the Crozet Islands. In South America, several small colonies have become established on Tierra del Fuego and even the Patagonian mainland. The birds prefer the level ground closer to the sea, occupying the beaches and snow-free valleys.

The King Penguin's breeding cycle is unusually long, and including the premolt, it can last up to sixteen months. This means the breeding colonies are constantly occupied, with chicks remaining in place throughout the year. The adults return irregularly during the winter period to feed the chicks, which may have to fast for several months between meals. No nest is built; instead the single egg is in-cubated on the parents' feet and transferred between birds on their return from extended fishing trips. Despite the fact that the breeding cycle is so long, breeding birds usually lay an egg every year; the chicks of late breeders are often too small to survive the winter, so King Penguins tend to breed successfully only every other year.

I was a very nerdy kid—the kind of kid who would rather spend hours looking at maps and atlases than watch TV. When we started travel-ing farther afield, there wasn't very much for me to do once the bird-watching part of our day was over, other than to read or play on my dad's phone—bear in mind that the variety of kids' apps in the early aughts was extremely limited. Dad had only one—an atlas app—that let you work out the distances between countries, oceans, and conti-nents. I "enjoyed" this app over many holidays, so it didn't come as a great surprise to my parents when, at age eight, I announced that I would visit each of the seven continents before I was fifteen. At the time, fifteen might as well have been thirty or fifty; in any case, it was years *and years* into the future.

Fast-forward to 2015: I was thirteen years old and had pretty much forgotten this childhood pledge when my parents announced we were going to Antarctica in December.

Having already visited six continents, a nerdy kid was about to see her fantasy come true. But it wasn't just *my* fantasy. Dad, who had lost his father suddenly, at fifty-three years of age, had also pledged to visit all seven by the time he was fifty. His father's early death, he told me, "focuses the mind."

Mum was on an even keel, her regimen of meds back on track, and she was as excited as I to spend ten days on board an icebreaker plow-ing the icy waters of Antarctica.

· · ·

There are about ten thousand bird species in the world. The more you travel, the more you will see, obviously, and as your list grows, the number of species you can add to it shrinks. It's a law of diminishing returns, and the remaining birds are usually the rarest, the hardest to see. If you're a rich and ambitious world birder, it's pretty much a no-brainer that you will use your funds to seek out these hard-to-find birds in distant locations: for example, a trip to the remote Pacific islands, where you might hope to catch a glimpse of a tiny handful of endemic birds.

From a birdwatcher's perspective, the number of birds in Antarctica isn't huge and was certainly not going to add volume to my list, but at the same time, I was hoping to see up to two dozen new birds. It is the quality and uniqueness of the experience that draws birders to Antarctica rather than the numbers.

A few days before Christmas, we flew to Chile and from there to the Falkland Islands, where we were due to set sail. It was strange to be greeted by a British customs official at Mount Pleasant airport; we literally couldn't have been much farther from the UK if we tried. While everyone else in our Antarctic party headed straight to the boat, we had decided to spend the afternoon birdwatching before boarding. We set off under a dark sky in a bleak landscape of grass and dwarf shrubs without a bird in sight. Arriving at Yorke Bay on the edge of the capital of the Falklands, Port Stanley, we followed the boardwalk through yellow flowering gorse bushes to a viewpoint overlooking a perfect crescent of silver sand protected by craggy cliffs. We could have been on a Scottish headland—that is, until we scanned the beach and noticed the huddle of Magellanic Penguins. Striped black and white, they looked like oversized humbug candies. A sign warned us off the beach because of the risk of unexploded land mines, a legacy of the UK's 1982 conflict with Argentina. Ironically, this exclusion has protected these important penguin rookeries and allowed them to expand.

The scene felt even less Scottish as I watched a pair of Kelp Geese feed among the seaweed on the tide line, the white plumage of the male a contrast to the finely lined dark brown of the female, but there was no sign in the bay of our target: the Falkland Steamer Duck. We needed a backup plan if we were to see our bird. So back in the car, we made a quick dash to a second beach, where we found the ducks surfing the waves just off the seashore.

Flecked grayish brown, the Falkland Steamer Ducks have a distinctive white stripe curling away from their eyes and a heavy yellow bill specially designed to break open shellfish. They are flightless birds, their wings reduced to the size of paddles, named for their habit of running across the water, wings thrashing like a paddle-wheel steamboat, and one of only two birds endemic to the islands. The Falkland Steamer Duck is a "must see" for any world birder; we were off to a great start. It is a large and particularly heavy duck, and notoriously aggressive. With most bird species, there is a lot of posturing, but not these guys. They have a bony knuckle pad on each wing and are prepared to use them. They will repeatedly hit any other duck foolish enough to enter their territory and have been known to beat smaller species to death.

We arrived at the ship, a small ex-Russian 1960s icebreaker, five minutes before the gangplank went up, the other eighty passengers already safely on board. Designed to clear a path in the water by pushing through the ice, this was a sturdy ship that, to me, screamed, *Adventure!* The stereotypical Cold War tech of buttons, levers, and blinking lights in the control room and whispers of it being an old spy ship only added to the general feeling that this was going to be an epic trip.

In all the world birding I had done up to that point, there were rarely other kids my age around, but now there were eight older American teenagers, all traveling with their parents. They seemed effortlessly cool in their trendy gear, and I was suddenly self-conscious. I was wearing very baggy thermal walking trousers, a couple of old jumpers, a woolly hat past its prime, and clumpy walking boots. I felt deeply *uncool*. And none of them were into birdwatching, so I felt even worse.

This was the first time there was a direct collision between my hobby and those who might tease me for it. I couldn't let that happen, so I put on my skinny jeans, hid my binoculars, and played cards with the teens, only to be shamefacedly dragged off by Mum and Dad when a target bird appeared in the skies.

It took me just a couple of days to get over myself. It became pretty clear, pretty fast that the Americans couldn't have cared less about what I did with my time on board and in fact wanted to hear all about my life in the Chew Valley. But I was thirteen, and for the rest of the trip, I tiptoed a fine line between feeling normal and being a bird-watcher.

The plan was to go island-hopping around the Antarctic Peninsula, but before we embarked on that leg of the journey, we made a stop at Sea Lion Island, on the southern edge of the archipelago, famous for its wildlife and its wilderness. The sun was high in the sky, and from the ship, the island's sandy beach with its acres of tussock grass gracing the coastline looked idyllic. A perfect summer's day, and it was actually *warm*. (Later, in Antarctica, preparing for any expedition on the land required at least half an hour to drag on snow boots, layers of thermals, and biosafety suits to ensure we didn't spread invasive species onto the islands we visited. I often felt like an astronaut about to be launched into space.)

Small rigid-hulled landing crafts, or Zodiacs, were launched from the ship. We would use the Zodiacs to get from ship to shore through-out our trip. Island bird populations around the world have been dev-astated by the accidental introduction of rats and mice, with the chicks of ground-nesting birds having zero defense against these predators. The Zodiacs were the safety zone.

As we approached the beach, we were greeted by huge, magnificent adult sea lions basking in the sun and sand. The males are particularly enormous, up to nine feet long, their shaggy manes as luscious as any lion's. They aggressively defend both their harem of much smaller females and their patch on the beach. It was the start of the pupping season, and the females, having carried their young for eleven months, would be pregnant again within days of giving birth—it struck me as a pretty relentless cycle of life.

Farther along the beach, we met Gentoo Penguins, completely fearless, inquisitively making their way toward us. They seemed to be asking, *Who the hell are* you *in the bright red parkas?* Perhaps they were twitching *us*.

If a child were to draw a penguin, the Gentoo is what they might come up with: white belly, jet-black head and wings, orange beak, and orange feet. These are the world's third-largest penguins, preferring the coasts and valleys to the ice. Unlike their Emperor Penguin cousins, Gentoo parents share the incubation of their eggs, and once hatched, both Mum and Dad forage for food in turns. The dune slacks beyond the beach were a sprawling mass of Gentoos, chattering loudly, unafraid of the soon-to-be Antarctic voyagers snapping pictures on their beach. The visit had been timed to see the chicks at their various stages of fluffiness. Among them were also many Magellanic Penguins and even a couple of King Penguins, the second largest of the penguin species behind the Emperor Penguin. The King Penguins looked like self-conscious teenagers whose growth spurt had kicked in before their peers; they stood hunched over, trying not to stand out in the crowd.

With their black heads and cheek patches of orange, these are fantastically striking birds whose numbers are in decline, a result of rising temperatures and habitat loss. They have been forced to travel farther away from their breeding grounds to source food. The fish they feed on, being sensitive to water temperature, are able to move toward the South Pole as seawaters warm, but King Penguins are tied to their subantarctic breeding grounds. They are already near the limit of how far they can travel, nearly 500 kilometers a week on fishing trips, but with climate change, they would need to add an extra 190 kilometers to reach the food they require to survive. This is a depressing thought—and none of this is their fault: it is all ours. The joy of birdwatching is too regularly overshadowed by the threat to habitat from both man and climate change. Staring at these birds, which were oblivious to the challenges they would have to face, I found it hard not to feel helpless and ashamed.

Back on board the icebreaker and sailing away from Sea Lion Island, we entered the treacherous Drake Passage. This notorious stretch of water, the gateway to Antarctica, lies between the southernmost tip of Chile's Cape Horn and the South Shetland Islands of Antarctica. The Atlantic, Pacific, and Southern Oceans meet right here, and without resistance from any landmass, as the icy waters from the south collide with warmer waters from the north, mixed with a strong wind—or worse, a storm—the turbulent upshot is an extremely uncomfortable journey.

There was no way around it. Forty-eight hours of seasickness and choppy waters beckoned. We had our pills, we had our tiny cabin, we could hide. I prepared to hunker down, picturing myself crawling along a heaving corridor, with a heaving stomach, to the communal bathroom. Everyone aboard felt the same.

Despite this gloomy prospect, I was still very excited. The Drake Passage is a landmark, a final frontier. It was that part at the bottom of the globe I had been staring at for years, the part that would complete all seven continents. Casting off from land to head south into these huge seas gave me a real sense of how small I was: a tiny figure in skinny jeans, on a tiny boat riding a vast expanse of ocean.

Normally Mum has only to look at the sea to become seasick. She had once started throwing up even before leaving the harbor on a seabirding trip with Dad before I was born, and had spent the next five hours hanging over the side of the boat, barely managing to raise her head every time a new bird was called out.

What was particularly scary about the Drake Passage was that there was no way back, no getting off the ship at the end of the day, no escape if the sickness kicked in. It would simply be a case of enduring it. Dad, on the other hand, considered himself a good sailor, and nothing was going to stop him from spotting the truly oceanic birds this stretch of water is famous for.

But the polar gods had smiled on us, and our journey through the passage was smooth. Afterward, the captain announced it was one of the smoothest crossings he had ever experienced. While no one was sick, the real win was watching the birdlife. If we weren't on deck, binoculars in hand, we were on the bridge, standing next to the "resident" ornithologist. At least, he was as resident as one can be on a ship. He had led more than twenty trips to the Antarctic and knew his birds. Other passengers had different levels of interest in the birdlife. Many had binoculars and certainly wanted to see albatrosses and penguins, but there were no hard-core birders, namely those who *needed* every species on offer. We were the only family running from one side of the ship to the other as birds flew from port to starboard or vice versa. The Americans had us down as British eccentrics, and they were probably right.

My highlight of the crossing was the albatrosses, the Black-browed Albatross bringing back happy memories of the Big Year. While it was a large bird, it was dwarfed by the Wandering Albatross that cruised alongside it. Now, this was a big bird! The largest of the two dozen species, it has a wingspan of up to three and a half meters: the wings locked into position, like a switchblade, by a specially adapted tendon, allowing it to soar effortlessly over the ocean waves and spend up to six years at sea.

Out on the ocean, it can be very difficult to gain a true sense of size and perspective because there is no frame of reference. So, it wasn't until a Black-bellied Storm Petrel joined the Wandering Albatross in the air that I was truly able to appreciate just how huge it was. With a wingspan of only forty-five centimeters, the petrel was *tiny* by comparison. With its pure white body, pink bill, and black-tipped wings, the albatross wheeled off across the stern and followed the wake of the icebreaker far into the distance. It turned into a vanishing speck and then was gone, all without a single flap.

Albatrosses spend most of their lives in flight, coming onto land only during mating season. Given that they can live for over seventy years, this is a staggering amount of time to stay in the air. Now another had appeared—the Gray-headed, ripping through the sky. I remembered one startling fact about this solitary creature: it holds the Guinness World Record for the world's fastest horizontal flier, reaching speeds of up to 127 kph. The longest period they spend on land is the four months following their hatching, after which they take to the air, where they will spend over half a decade at sea before returning to their original colony to breed. But just like the King Penguins, their population numbers are decreasing.

As temperatures rise in the sea, fish stocks tend to move around, and the once-reliable destinations for food become depleted. If an animal has to walk an extra twenty or thirty kilometers to feed, it is out in the open for longer, facing not only exhaustion but danger besides. This prolonged absence and increased vigilance are so all-consuming that its chicks might become neglected. It takes a long time for a species to adapt to new breeding and feeding grounds, and if an animal is already struggling to feed, a further caloric reduction can push a species over the edge toward extinction.

As well as suffering food shortages because of climate change, these albatrosses are also the victims of longline fishing, diving for the hooked bait, thinking it is an easy meal before becoming trapped underwater and drowning. What a sad, ignominious death for such a magnificent bird.

As Dad explained this bleak reality to me on the deck of the ice-breaker, the sun high in the sky, I thought about freedom, the trait that I, and most birders, believe is one of the most compelling characteristics of birdlife: a bird's ability to go where it will. I was learning more and more that this freedom is conditional—it is regulated by humans, habitat destruction, and global warming. As long as these albatrosses stayed in the air, they were safe, but this is not a realistic or natural option. They have to feed and they have to mate. Sobering as it was to learn that maybe one day in the future, as species become extinct, world bird lists wouldn't feature the abundance we enjoy today, something sparked inside me. Watching these giant birds ride the thermals, indifferent to human concerns, I realized it was people like me and Dad and Mum who could talk about the devastation caused by human actions. Surely, along with the privilege of visiting Antarctica, wasn't it our *duty* to promote and protect the poles?

I mulled this over as Slender-billed Prions gave way to Antarctic Prions toward the end of our journey through the passage. The latter were minuscule silvery-gray birds with an inverted black *W* stamped across their wings and back. How could such a tiny species survive in this hostile environment? I wondered.

There was a final salute from a passing Southern Royal Albatross, almost as big as its Wandering cousin, before we reached the colder waters of Antarctica and huge lumps of ice started to appear in the surrounding sea: our first icebergs. I felt as if I were in a movie.

While whales and other sea creatures aren't by any means absent from the waters of the Drake Passage, it wasn't until we left this stretch of water that the sea began to writhe with fish. It was shallower here and teeming with food. Humpback, fin, and sperm whales circled the ship, happy to be around us, the days when they were almost hunted to extinction behind them.

From the Drake Passage, we sailed toward the rocky, mountainous landscape of Elephant Island, still over two hundred kilometers from the Antarctic Peninsula. The island is also where, in 1916, the polar explorer Ernest Shackleton and his crew of twenty-seven sheltered following the loss of their ship, *Endurance*, after it was trapped in the ice in the Weddell Sea. From there, Shackleton took a lifeboat on a sixteen-day journey to South Georgia to seek rescue for his men.

Looking across the claustrophobic bay on Elephant Island, where the stranded men had once fashioned shelter using the two remaining lifeboats, I pictured the lives of these pioneers, stranded for four and a half months, struggling to survive in a bleak environment with little food—the penguins had migrated north on the day they landed— while they waited for Shackleton to return. Day after day, they readied themselves for rescue, only to be disappointed. The hopelessness of their situation felt real to me now as I gazed into the endless open sea, silent and empty.

We headed farther south, experiencing another Attenborough moment as we wound around a rocky outcrop and icebergs as tall as New York City skyscrapers surrounded us. I could almost hear his voice, marveling at the size of the glaciers. I could also hear the ice singing as we approached the towering white hills. Horrified and fascinated, I watched a pod of orcas separate a whale calf from its mother, chase it down, and eventually drag and drown it in the depths of the Antarctic Ocean. A leopard seal caught a fleeing penguin, thrashing it repeatedly from side to side on the surface of the sea, skinning it alive to remove the tough, feather-covered skin before gorging on the meat and fat beneath. Whenever we visited the vast penguin rookeries, there were the ever-present menacing Brown Skuas patrolling the colonies, ready to snatch the cutest, fluffiest chicks from any unattended nest. This really was nature at its rawest. It was harsh but at the same time so much more honest than the row upon row of prepackaged meat in my local supermarket. These predators also had their own young to feed.

Having survived the Drake Passage, Dad and I were about to risk our lives once more. When Mum had booked the trip, she announced that we would spend one night camping on the ice. She had imagined tents, mattresses, and lots of bedding, more ice *glamping* than ice camping. When it transpired that, no, we wouldn't have tents and blow-up beds, but would be sleeping *on the ice* in bivvy sacs (essentially, waterproof sleeping bags), she point-blank refused. "I'd do it for an Emperor Penguin," she told me. "But not for *fun!*"

We woke to clear skies and the ever-present white panorama on Christmas Eve. There are lots of lyrical and exciting descriptions of the landscape of Antarctica, from *Endurance: Shackleton's Incredible Voyage* by Alfred Lansing to Beryl Bainbridge's *The Birthday Boys*. I'm not going to add very much to these passionate accounts. In any case, for me, this had become a journey of the senses, which *my* words only seem to diminish. The silent beauty of my surroundings wound its way into my body; I could taste the pure white of the looming glaciers, smell the miles of cloudless skies. The vividest color in the whole of Antarctica is the blue of the ice. It is a special shade of *spectacular* that I have only ever seen at the poles, hidden away like a luminescent jewel in the depths of the icebergs. Out on the water in the Zodiacs, surrounded by the towering, snow-covered rocks of the islands, I felt like an explorer in a landscape in which humans are the aliens.

But it was the blue that drew my eye, again and again. When the gray husks of ice cracked open, they revealed a color lit from within, so intense, so unique, that it was hard to believe it exists in nature.

Earlier that year, in the spring, I was on the set of the TV show *Springwatch* and was chatting with naturalist and presenter Chris Packham about the iconic birds we had seen around the world. I had seen every bird he threw at me until, finally, he said, "I bet you haven't seen a Snow Petrel." I had to concede defeat. This beautiful bird is found only in Antarctica.

"I bet I'll see one when I go in late December," I told him.

"Really?" he said. "I'll be going in early December! Maybe we'll both get lucky."

The game was on.

Christmas Eve afternoon, we were navigating the ice floes in the Zodiacs, moving toward the icebergs; they were tall enough, even without stopping to consider the fact that 90 percent of each one lies below sea level. A black dot streaking down the side of a white pillar mountain caught my eye. I hung on to it. It was the eye of a bird, but *which bird*? Camouflaged by the backdrop of pure white ice, only the Snow Petrel's beak and eyes had been visible; now it flew against the blue skies. No larger than a London pigeon, these angelic-looking birds, which feed mainly on fish, think nothing of polishing off seal placentas and dead penguin chicks!

I was planning to email Chris as soon as we got a signal on the boat, but once the internet loaded, I discovered that he had seen his Snow Petrel two weeks before me—and on the same boat we were traveling on now!

He had won this round, but I didn't mind too much. The Snow Petrel had been my early Christmas present, but there was more to come when I was handed the best festive gift I have ever received. An urgent message came over the PA of the icebreaker, inviting everyone on board up to the deck. What we didn't know was that our sister ship had passed by a Chilean Antarctic base two days previously and sighted an extraordinary bird. They had tipped off our captain, who had quietly diverted our ship to the base. Was this to be my ultimate twitch? There, on the ice ahead of us, was a young Emperor Penguin.

The news had come through just as we sat down to lunch. Abandoning our meals, Mum, Dad, and I were the first to arrive on deck—surprising even the resident bird guide with our haste. To most people on board, it was just another penguin, but we were going insane.

Emperor Penguins breed a long way inland, deep in the Antarctic interior, and the only guaranteed way to see them is to take a dedicated helicopter flight to a breeding colony. But here it was, standing over a meter tall. This slightly scruffy specimen was, in penguin terms, a teenager: old enough to have left its parents but not yet old enough to breed. Perhaps it was on a penguin gap year, visiting the various Antarctic bases, gathering stories to tell its own kids in future years. I could have stared at it forever, but we had to get to our sleepover on the ice.

Round two to me, Chris Packham.

After a "last supper" aboard the ship, those of us reckless enough to bed down in the Antarctic snow on Christmas Eve took the Zodiacs to Leith Cove, where we were each handed a shovel on landing and told to dig our sleeping berths. Thank God Mum had decided to stay behind. Sleeping on the ice was one thing, but this felt a bit like digging your own grave!

Keen to put off bedtime, we campers huddled together to watch the sunset. Summer in Antarctica enjoys six months of daylight, and the sun never completely disappears behind the glacial mountains; the sky, while certainly darker, glowed a dull russet orange at the horizon.

Later, in layers of thermals and sweaters, I climbed into my all-weather sleeping bag and enjoyed one of the best night's sleep of my thirteen years. I slept so well, in fact, that Dad had to wake me the next morning. As we packed up, I thought about the stranded crew of the *Endurance* once more. I would shortly be back on board the icebreaker with a hot breakfast and welcoming shipmates. Shackleton's men had had to repeat my night's "adventure" again and again and again, until surely some part of them must have begun to feel that their wilderness stopover was nothing more than a cemetery.

Despite my glowing reports of a warm and comfortable night, Mum was unrepentant, declaring she was glad she hadn't camped on the ice. She couldn't be "doing with teenagers talking all night long" while she froze in the snow.

So what do you do on Christmas Day in Antarctica? By this point, six days into the trip, I had become firm friends with the Americans; we had bonded over card games, but the activity that really sealed the deal was a subversive Antarctic tradition. Dressed only in a swimming costume, participants are dared to jump into the icy sea, and *stay there* for as long as possible. Forbidden to do this from the ship, we repeated the challenge by diving into the ship's pool, filled with the same icy seawater. Who could jump in the most times? Who could stay in the longest?

I wasn't going to be beaten. Nerdy stubbornness kicked in, and despite my fear of *dying* as I crashed into subzero temperatures, my flesh instantly punctured by icy needles, I stayed put. Ten seconds, twenty, thirty...my mind was on only one thing: to remain here as long as I could without passing out. Again and again we jumped in, until there was only me and one other boy in the competition. Hypothermia wasn't going to stop me. Or him, apparently. We drew first place.

. . .

We had crossed the Antarctic Circle, cruised the ice floes in Zodiacs, landed on the islands, even camped on ice, but for purists, there was still one last thing to do: set foot on the Antarctic Peninsula itself. This is never guaranteed—"all activities are weather-dependent," as the brochure likes to point out. December 27 was our day of reckoning, and we awoke to a swell on the water. Not good. The ship sat offshore while we nervously ate breakfast—once again everyone was tense. Then the shout went up: the captain had given the go-ahead. We pulled on our cold-weather gear and the Zodiacs were dropped into the water.

We landed on the gray, pebbly beach of Brown Bluff, with its huge, scattered boulders and ice-covered slopes of scree, backed by towering brownstone cliffs, where Snow Petrels breed among the misty tops. This was some scenery! There was a raucous beach party of the three commonest species of penguin: Gentoo, Chinstrap, and Adélie. The first thing I noticed about these massive colonies was the smell—they absolutely stank of rotting fish—and the second was the noise. They love to chat, and I guess this was their Christmas party.

With no land predators to concern them, they were oblivious to any danger we might present. Out at sea they were far more wary, porpoising their way toward land like oversized flying fish. One minute they were under the water, the next they were launching themselves through the air. We watched seething colonies flop onto land—hundreds of thousands of penguins, as far as the eye could see. They began to march, following invisible paths, as if they were heading toward a mass demonstration.

We climbed into the Zodiacs and sped back to the ship. Turning around to take a final look at the undulating black mass of penguin life, I tried to fix the picture in my mind. It would be a sight to savor, a final glimpse of life in Antarctica as we began our long journey home.

It was hard to imagine going back to the Chew Valley and returning to school. This had been a short trip, but somehow I had convinced myself it would never end, or that I had always lived on the icebreaker, always traveled by Zodiac. It was a small insight into how Mum feels when she's away, as though the outside world just ceases to exist when there are birds to be found and adventures to be had. Her ability to live in the moment is her superpower, and given that her other powers aren't so reliable, it's been a blessing to my family that we can usually make the most of the time we have away.

I now wonder whether Mum's goat cheese baguette an hour before we flew home was a subconscious desire to stay in Antarctica. Given that one of her daily meds is contraindicated with cheese, and one of the possible side effects is a stroke, it was a poor choice. A minute into the flight, as we were still gaining height, Mum started complaining about feeling too hot. Dad took her pulse and shook his head slowly. "Your pulse is racing. You need to breathe slowly." While it was a panic attack and not a stroke, she still felt deeply uncomfortable. Dad did his best, trying to calm her down with relaxation techniques; only once did he have to physically restrain her. "No, you can't get up—we're still climbing and the seat belt sign is on!"

We made it home in one piece, and cheese is now off the menu.

I had visited my final continent, the Antarctic. I had camped on the ice, swum in its freezing waters, and seen the bird of my dreams, a Snow Petrel. As usual, nothing had changed at home. I returned to my small village; school and lessons and everything felt a little ... anticlimactic.

I was changing and beginning to understand that while my two worlds felt incompatible and rarely crossed over, they would have to accommodate each other because they were both part of *me*. For the first time, I began to appreciate the stability of routine. I liked my friends and enjoyed school. Camp Avalon had given me the opportunity to make a small difference, and I was ready to do more, to campaign more actively to stop biodiversity loss and climate change while promoting global climate justice and equal access to nature for all. I was also beginning to see our trips as not only a break for Mum but a period during which each of us could build a bit of resistance to whatever life decided to throw at us next.

10

California Dreamin'

CALIFORNIA CONDOR

*The California Condor, the largest North American land bird, is
a critically endangered New World vulture. Long-lived but slow
to reproduce, this condor can live for sixty years. It does not start
breeding until the age of six, however, and then raises only a single
chick each year.*

*In 1987, the remaining twenty-two wild birds were taken into
captivity to form part of a breeding program, with the ambition to
return the birds to the wild when conditions were more favorable.
With various threats to the wild population, there was a very high
chance that, without intervention, the species would become extinct.
The program has been successful, and the current population stands in
excess of five hundred birds. This iconic species is deeply significant to*

many Native American groups and plays an integral role in several of their traditional stories.

Instead of pictures of albatrosses and Snow Petrels, I posted photos of penguins on my private Instagram, and my friends responded with "Coooooooooool!!!!!!" and "OMG. Looks amazing. Sooooo jell!"

It was becoming easier to talk about these trips because I avoided all mention of birds. There was nothing geeky about hanging out with penguins, but in the summer of 2016, I would be taking a trip my peers might be more than half interested in.

California, but more significantly, America.

We had all grown up with American films and TV, from *The Muppet Show* to *Breakfast at Tiffany's*, from *Buffy* to *Gossip Girl*. I wanted to see a yellow cab, eat in a diner, and lose myself in city streets overshadowed by soaring skylines. There would be no hiking in cloud forests or plowing a 4x4 through a dusty national park with one eye trained on the landscape for prowling lions; this was the United States. When I told my friends, however, they seemed just as uninterested in this trip as they had been before. Of course! My peers weren't judging me for the places I went to and what I got up to; they simply weren't all that bothered about what *any* of us did over the summer.

My reluctance to publicly and fully embrace my inner anorak was not the only thing stressing me out in 2016; I was also coming up against racist abuse on Twitter. By this point, I had thirty thousand Facebook followers, and my Twitter had a dedicated ten thousand. My *Birdgirl* blog was thriving, having received around two million views. Donald Trump was running for the White House, and Brexit was on the horizon—both events opened the floodgates to an altogether more hostile online environment. Suddenly, those who disagreed with your politics thought nothing of hurling insults or threatening your life. An increasing number of Americans had begun to respond to my tweets whenever I talked about racism or, more specifically, referred to the failure of the nature sector to be more inclusive. None of my content was extreme, but the responses were frightening.

It is in hindsight that I now recognize I was the perfect target: a fourteen-year-old schoolgirl, relatively new to the platform, with few followers to defend me. The Islamophobic, sexist, and belittling comments were often difficult to ignore. Initially, I was hurt by the hate: I was no stranger to the casual racism that haunts a classroom, but this was different; it was toxic.

There were two ways to tackle the abuse, and I considered each one carefully. I could shut down my social media, which was tempting—it would certainly and immediately help me sleep better, but I also really loved Twitter. Enabling me to connect with not only like-minded birders but also a growing population of climate and race activists, sharing ideas and information, it was a platform that felt immediate and immersive. Why should I be hounded off by people I didn't know and was never likely to meet? The second option was to tackle it head-on, and this is the path I chose. I began to mute or block abusers, and posts featuring incendiary words were automatically hidden. I didn't have to be the bigger person here. I learned a valuable lesson when I had, on occasion, confronted a racist on my feed and soon realized that there was little to no point: they weren't there to talk *to* me; they were there to harass anyone who challenged their narrow view of the world. It just wasn't worth it. I didn't need people to agree with everything I posted. While I was happy to debate issues around inclusivity in the nature sector, I wasn't going to tolerate being told by some random bigot, British or American, to "go back home to Bangladesh" and sort out their problems first, or how all Muslims "should be killed."

. . .

America's Golden State is a birders' paradise, with its diverse habitats and strong emphasis on conservation. I would switch off the twitter of the internet for a bit and listen instead for the tweets of the 650 bird species recorded in California.

I was excited for the birds—of course I was—but this was *America*, a country so much bigger than its geographical size, so much more powerful and influential than any other landmass on earth. I wondered, not for the first time, how it could ever live up to my imagination. But first, I had work to do—pressing work, an ambition that was as challenging and all-consuming as birdwatching.

The previous year, after the inaugural Camp Avalon, I had written to various nature organizations, hoping to talk to them about diversity and inclusion in the natural environment and specifically to query their initiatives in this area.

I was now at an age where the idea of *systemic* failure was beginning to make a lot of sense to me. It was becoming clear that in order to tackle inequality, I needed to talk to the *bosses*, those in charge of the nature sector, because whatever they were doing, it wasn't enough to encourage minority engagement. My camp was a great initiative, sure, but nothing was going to change on a national scale if the organizations and charities who worked solely in the area of nature and conservation did not invest in or embrace diversity.

The organizations had all responded positively to my camp. Many asked to meet up so we might discuss my ideas with them. My ideas? I was fourteen years old; I had zero ideas about how they should work on their engagement strategies. Surely there were experts for this stuff?

I was introduced to the countryside at an early age by parents who believed in the merits of wandering the hills, valleys, and lakes of the Chew Valley: I was lucky. For my family, the outdoors wasn't just a place beyond our neighborhood, or a stretch of empty land you drive by on the way to somewhere else. I also know that camping, birdwatching, and hiking aren't for everyone, but I don't accept that specifically VME people should be denied the chance to decide for themselves. And they must have been denied at some point, because where were they? This was the foundation of my thinking: include *everyone*, and some might want to stick around.

But kudos to the nature sector; they wanted to talk, and if I was enjoying these conversations, then surely these CEOs might feel the same sort of inspiration that drove my camp. I wasn't going to be able to tell them what they needed to do—I needed help. I showed Mum the responses I had received. Wouldn't it be better if I could talk to them all at the same time? Mum, ever since her solicitor days, had kept her finger on the pulse of local activism. If anyone had a clue about next steps, it was her.

"Maybe a conference? You would need to find speakers, *and* somewhere to hold it."

And so, Race Equality in Nature was born: a conference for members of the nature sector with speakers who knew more than I did about implementing policy.

It was a good idea, and it would have just stayed an idea without Mum's help.

She began to approach local race activists for their advice. She was as invested as I when it came to equality in nature, having endured many years before my birth as the only Bangla woman in the local birdwatching community. She wanted the conference to inspire leaders to think about the people who were shut out, to help them visualize what a new approach might look like. In their youth, my mum and her sisters were active anti-racists, well versed in the lingo of exclusion, diversity, and most significant, *action.*

Mum pointed out that the makeup of those running these companies was similar to those she encountered in her early days as a lawyer: left-leaning, liberal types who eschewed racism. She understood them very well, and for all their fair-minded politics, they weren't doing much to help.

While we all believed there wasn't an active campaign to withhold nature from VME people, it was also true that there was a lot of work to do when it came to convincing leaders that sitting back and doing nothing was not an option.

My mum's oldest sister—my *khala*, or aunt, Monira—is a leader in the field of race and diversity. She is now responsible for diversity within a large NHS trust. Her speech would underline the point that an exclusionary attitude within the nature sector, however unintended, has the potential to limit a child's choices, ambition, and health.

My *khala* Lily would speak too. She set up Bristol Multi-Faith Forum and works with organizations such as the NHS organ donation unit to enable them to connect with marginalized minority ethnic communities. She understood that for many minority ethnic communities, there was a close link between race and faith. Her work resonated with me because I saw nature as a unifying force, regardless of religion or political beliefs.

And Mum, of course, had been a solicitor and partner in her firm. By the time she qualified as a solicitor, she was on the Bristol Law Society Equality, Diversity & Inclusion Committee.

Together, the sisters were a powerhouse.

I had briefly met the television presenter, writer, and conservationist Bill Oddie, a birder since childhood, the year before, at the Bristol Festival of Nature, and I approached him. Would he talk at the conference? Yes, he said, he would. Ever since I read his memoir, *One Flew into the Cuckoo's Egg*, I had felt more than a birdwatching connection with him.

Like Mum, he, too, suffered from crippling bouts of depression and mania. His mother had also been sectioned when he was a boy, and she stayed in an asylum for nine years. Just like our family, he had found some respite in nature. But Mum also found respite in having a mission, and now her mission was the conference. She raised sponsorship money, reached out to strangers for advice, drew up lists of attendees. Was it mania? I asked Dad. No, he thought it was just good old-fashioned zeal.

Dad ensured that the event ran smoothly, locking down details, checking everyone was where they needed to be. He was good at this stuff because he had to be; his organizational skills reflected pretty much how he managed family life. On the day, he also filmed the whole conference, the only surefire way of avoiding being on camera himself, he said.

I was, of course, terrified it would be a disaster. I imagined myself in a vast hall, sitting onstage alongside the speakers—busy people who had taken time out of their day—waiting for no one to show up. My invites had mistakenly ended up in the spam folders of many participants, which accounted for the initial and dismal lack of response, but Mum's follow-up ring-around saved us from an empty conference room.

Almost one hundred people attended, including representatives from the BBC, the National Trust, and the Heritage Lottery Fund, but there were no CEOs—therefore, no decision- or policymakers. Nevertheless, a welcome speech had to be made. The room fell silent as I took the podium, cleared my throat, and began to talk about birdwatching in a community of mainly White men, where I rarely crossed paths with anyone who looked like me. I talked about my camps, where VME teenagers not only connected with nature but actively enjoyed it and wanted to come back. Efforts on the part of the nature sector had to improve, and together we needed to identify and overcome the barriers to VME people going out into the countryside. Barriers such as poverty, feelings of exclusion, and fear of hate crimes. The sector's blanket "you're all welcome" invitation wasn't going to cut it. They were going to have to reach out to VME people in their own spaces and places.

Around a campfire at the end of a successful camp weekend, we would discuss how being outside in nature made us feel, both physically and emotionally. While flames flickered in the dark, we would talk about racism, poverty, issues of identity, how none of these hurdles was going to disappear overnight, but that by engaging with nature, by relishing the fresh air, we were gaining tools to support us and help us cope with everything else life threw at us. This is what was missing from the nature sector: the simple message that nature is good for everyone, but especially those in need.

Over the course of the day, topics ranged from fear of hate crimes to socioeconomic inequality, from general societal racism to mental health awareness. The audience was animated and engaged with the speakers. It was undoubtedly an inspiring day. Change was in the air, and I began to feel a little glow of hope that in the next couple of years I might not be the sole Bangladeshi face within a group of mainly male, mainly White birdwatchers.

I flew into San Francisco feeling lighter. Things were looking positive—we had ignited a conversation. The conference was a brief glimpse of what was possible if you just made a move in the right direction: someone, sooner or later, would walk toward you.

Wildfires were still raging in California when we picked up our vehicle, and we had to abandon plans to travel farther south. By the end of 2016, over seven thousand fires would have consumed more than 2,500 square kilometers of forest and killed upward of sixty-two million trees. The obvious tragedy for birds and birders was the loss of habitat. These fires were out of control, but in themselves forest fires aren't inherently bad.

In the past, people knew how to manage their forest fires. Such controlled fires are often part of the natural cycle of the life of a forest, clearing dead trees and preparing the land for new growth. Increasingly, these controlled burns are prohibited, but fires rage nonetheless—only now there is no controlling them. The lack of rain and rising temperatures in the summer months pretty much guaranteed that all that was needed was a single spark to launch the domino effect of devastation for the Californian woodlands. Never before had the repercussions of climate change felt so immediate or visible to me. It was an uncomfortable experience; here I was on holiday while homes, forests, and habitats burned.

Peeling away from San Francisco, our first stop was Anthony Chabot Regional Park. Featuring a lake, meadows, and woodlands of eucalyptus and oak trees, it made me feel as if I had stepped out of a fiery apocalypse and into a birders' paradise, a jarring transition.

We were driving in the hot sun along rivers and through pine forests, through giant redwood forests in which we felt Lilliputian; everything really was bigger in America. There were signs advising tourists not to feed the squirrels, owing to cases of bubonic plague among the rodent population in the area. This isn't the fourteenth century, I reasoned, and should we somehow, maybe by accident, catch the plague, antibiotics would sort us out pretty fast. All the same, when we saw a lone squirrel on a log at the foot of a eucalyptus tree, pawing the air as it gnashed its teeth, Dad sped up—we weren't about to take our chances despite the leaps and bounds of modern medicine.

Our first wave of birds included the Calliope Hummingbird, new for me and a rare prize for my hummingbird list. This tiny creature, the smallest breeding bird in America, doesn't wear the same full palette of bright colors as its Ecuadorean cousins, but it was just as pretty. Only seven centimeters long, they migrate as far south as Guatemala and Belize for the winter—astonishing. The diminutive male has a soft pale-gray underbelly and a beard of magenta rays from its bill to its throat. This was another bird we spotted on the side of the road where we had parked up to eat our sandwiches. Once again, the wonders of roadside verges!

A flock of American Bushtits followed us through the park; rotund and very chatty, they could have been fuzzy brown tennis balls swooping through the air. But it was the adult male House Finch that really made an impression. Streaked with brown-and-white feathers, the male House Finch looks as though someone, as an afterthought, decided his complexion was a little pale and splashed him with a rosy-red powder to give his head, neck, and breast a flush. The red coloration comes from pigment in the berries and fruits House Finches eat: the higher the concentration of pigment, the redder the male and the greater the chances of being selected by a female as a capable breeding partner able to help with feeding their chicks. Males with lower pigment appear more orange or even yellow. Originally from the western United States, they were eventually released in New York City after a failed attempt in the 1940s to sell them in the caged-bird market as "Hollywood Finches." They began to successfully breed in the wild and, fifty years later, had spread across the eastern United States and into southern Canada.

We'd forgotten to bring any CDs, so after leaving the park, we made a quick stop to pick up some music. After the usual debate about musical taste between my parents—Mum loves country music, Dad does not—the soundtrack to the rest of our Californian tour was decided: Nirvana and Loretta Lynn, on a steadily irritating loop.

Debi Shearwater is famous for three reasons: The first is her extensive and impressive knowledge of seabirds, and the second is that she is the founder of Shearwater Journeys, which offers seabirding expeditions off the Californian coast on her famous pelagic boat rides from Monterey Bay. The third reason is that in the birding movie *The Big Year*, Anjelica Huston's character, Annie Auklet, is modeled on Debi.

We were running late—the boat was ready to leave just before dawn, but we were still parking the car, far from the quayside. Dad sped ahead, barking at me and Mum to stop dawdling and hurry up. "This isn't a walk in the bloody park, we need to run! They're not going to wait for us." But he was wrong: they were waiting, and they didn't look very happy about it.

It's miserable being stuck on a boat in the driving rain, and even though the outlook was good, I knew better than to trust clear skies and had had the foresight to don a couple of extra layers. The idea of being cold *and* feeling sick was too much. And I had been right to be wary.

As soon as the boat left the bay and sailed into the choppy early-morning waters, Mum and I were immediately and violently seasick over the side. While my stomach heaved, I cast an eye over the gray waves… There was something lurking beneath the surface. But there was no time—or inclination at this point, to be honest—to investigate further.

So much for "sunny California." Today the weather was terrible, the sky a heavy gray canopy over the thrashing surf. The clouds, thoroughly tired of hovering aimlessly, cracked open to release a waterfall of rain. There were about thirty people on board: a tight fit, but no birder worth their salt would have sheltered in the admittedly tiny ship's cabin. We were a boatload of obsessive birdwatchers and would do whatever it took to spot the maximum number of birds.

Looking around, clocking my fellow passengers, I might just as easily have been aboard a UK pelagic: there wasn't a single VME face on the boat other than mine and Mum's, just the usual suspects with stoic, wet faces. But that tingle of hope I felt after my conference briefly ignited somewhere inside my rib cage: change was coming.

An hour later, the weather had eased up, as had my nausea. The misery passed. It was no longer raining, and bolts of blue tore through the gray skies, as did a Black-footed Albatross. My hair whipping in the wind, I raised my binoculars. Although large compared with other seabirds, the Black-footed Albatross is the most compact member of the albatross family. Its wings open and extended to their full seven feet, looking like a U-2 spy plane, it swooped into the waves to scoop up the flying fish.

Moments later, silky black, with a bright blue patch at its throat, a Brandt's Cormorant bobbed into view among the waves, disappearing under the surface as it dived deep into the waters of the bay for food.

Shearwaters are so named because their stiff wings seem to shear the surface of the waves as they fly. We saw two species that day: the Sooty and the Pink-footed. Holding the record for the longest migration ever recorded is the Sooty Shearwater, covering sixty-four thousand kilometers in a single year. While they're not the most colorful birds, the way they glide so smoothly over the water makes up for their less-than-exciting plumage. This species was the first Debi had seen on her maiden pelagic trip, and she had been so dazzled by the birds that she changed her surname to theirs.

And Debi herself was a huge presence, striding over the wet decks, yelling, "Off to the left!" or "Just coming out of the clouds!" while pointing at patches of dark gray in the sky. Once or twice she took the boat off course, much to the ire of the more blinkered birdwatchers among us, insisting we look away from the skies and into the sloshing waves, where whales lurked. I was delighted. I have always loved seeing whales.

The clouds finally parted to reveal a gorgeous sun, and the dark mass beneath the waves I had spotted earlier reappeared. Titanic, menacing, keeping pace alongside the boat, was a blue whale. I felt sure that this huge mammal could have flipped us over with one swift strike of its tail. Its massive back, hovering just beneath the waves, looked like a tiny island in the sea; it was radiating the color blue through the water.

There was no way to tell how big it was. You could see some of it, like an iceberg, but you also knew that most of it was beneath the water.

Poised on the rocks as we sailed back to the docks, a couple of Black Oystercatchers watched the boat. Brown-black save for their red-ringed yellow eyes, red bills, and pink legs, these startling birds began to hammer away at the crabs they had just grabbed from the shoreline. As the pale flesh hung in their bills for a moment before being thrown down their throats, my stomach briefly heaved. It was good to reach dry land, however wobbly I felt.

Maybe staring intently into the sky doesn't sound so tiring, but the drenching from the rain, the bump and grind of the waves, and the seasickness were exhausting. "Sandwiches?" asked Dad, pulling foil-wrapped parcels from his bag. Mum and I shook our heads—we were barely able to walk in a straight line, never mind eat.

All Dad really wanted for his birthday was to glimpse the critically endangered California Condor, the largest of the North American land birds. This condor nearly became extinct before 1987, when the California Condor Recovery Plan captured the remaining twenty-two wild birds to be placed into a captive breeding program. Their declining numbers had been driven by the impact of DDT, a pesticide that thinned the eggshells of large birds of prey, causing them to fail. This was infamously highlighted in the 1962 classic *Silent Spring* by Rachel Carson, a book that is widely regarded as having ignited the modern-day environmental movement. The decline was exacerbated by the poisoning of birds feeding on carcasses of animals killed with lead shot, poaching, and habitat degradation. The recovery plan was successful, and in the early 1990s, captive condors were released into two separate populations in California and in Arizona. Today the world population of the California Condor is just over five hundred, but the bird remains critically endangered.

Since the early 2000s, the Indigenous members of the Yurok Tribe have actively engaged in the restoration of the rivers, forests, and prairies of their ancestral territories in Redwood National Park in Northern California in preparation for the return of the California Condor to their lands. The condor plays a central role in Yurok spiritual and cultural beliefs—it is a sacred animal and intrinsic to their obligation to heal the world, with its feathers used and its songs sung during their World Renewal Ceremony. The tribe also considers the condor a critical part of the ecosystem. Using fierce talons to tear open the tough hides of grizzly bears, the California Condor invites other scavengers—such as raccoons, ravens, and skunks—to feast.

Inspiring projects such as this, driven by local Indigenous communities, are for me conservation at its best, combining people, wildlife, science, and culture to reinstate a natural harmony to the world.

Twenty-three million years ago, the eruption of multiple volcanoes molded the unique landscape of what was to become Pinnacles National Park. East of the Salinas Valley in Central California, the park is twenty-six thousand acres of otherworldly rock formations, as well as spectacular woodlands and chaparral (thickets of shrubby bushes), and the perfect habitat for skulking birds. Pinnacles is also a nesting place for the California Condor. We had now entered the hottest days of the trip, and today's temperature—forty degrees Celsius—created a rich stew of heat and aridity.

If I were a bird, I would have taken shelter at this point, folding my wings in surrender within the branches of a giant oak tree. Thankfully, birds don't think like me, because if they did, we would never have seen the condor. Trekking slowly up a steep rocky incline, sweating and cursing, we made our way to the peak. "It will be worth it," Dad reassured us. "We'll be glad we made the effort." He was usually right about this stuff: Hadn't he, after all, propelled me through my first Big Year and urged me to sleep on the ice in Antarctica?

The air started to boil, heat hazes fanning the horizon in all directions. I wanted to peel off my skin, and thought longingly of my "bed" in the ice, but all notions of escape were instantly vanquished the moment Dad, who had set up the telescope while Mum and I were complaining about the heat, beckoned us over. Four adult California Condors were circling the skies.

These birds, very simply, filled the sky. They were striking, magnificent, but also scary as hell. The condor's head is fleshy and featherless, a thick black ruff around its neck. They looked like an oil painting of a group of old Elizabethan men, Gothic creatures through and through. Their plumage of pure black is highlighted by a large white triangular patch on the underside of their wings, which extend up to three meters, spreading out into long feathered "fingers" at the tips. I shivered, recalling that their favorite food comes in the form of pig, cattle, and deer carcasses.

"Thank you," said Dad to no one in particular. "That's what I call a birthday present."

"And sharing is caring," added Mum, giving him a small shove to take her turn at the telescope. Despite the heat, we stayed put until the condors were ready to leave, drifting away on their vast, cloaklike black wings.

There was a period not so long ago when no California Condors flew overhead. Today, they are back, and as of 2019—when a statewide ban on lead ammunition was issued—little stands in the way of the condor's full recovery to the skies.

Yosemite is one of America's most iconic national parks. Home to waterfalls, giant meadows, and ancient sequoia groves, it is also famous for the towering granite monoliths of Half Dome and El Capitan. They were impressive, certainly, but I didn't fancy climbing either of them. One of our target species was the Mountain Bluebird, and so began a three-day campaign, hounding hill and valley in search of this reclusive creature.

We had only a few short hours left in Yosemite—our parks permit was running out—when we came across a couple of rangers at our final stop. Never one to hold back, Mum asked after the bluebird. They smiled and casually pointed to the crest of a low hill. We walked down toward the lone, dead tree at the bottom of the slope, and only a little way along the track, in stark, late-afternoon light, two Mountain Bluebirds were flitting among the dead branches. Compact and round, with heads and wings of bright blue feathers, they had fat breasts that were dusted with a light blue plumage reminiscent of the Antarctic ice. I felt like the luckiest girl alive as I watched them dancing in the air. We had seen plenty of other birds during our time in the park, but at the back of my mind was the mysterious Mountain Bluebird. It's always that way when you're looking for a target bird: your thirst isn't quenched until it's quenched. The sky was a deep, dark pink as we drove away from Yosemite.

The tiny town of Lone Pine sits in the vast Owens Valley, the Sierra Nevada to the west. We had been alerted by our birding app, eBird, that a Zone-tailed Hawk had been seen among the kettles of vultures there. The hawk was a smart guy: he had mimicked the appearance of the vultures, allowing him to make sneak attacks from within their midst on his unsuspecting prey of lizards and small mammals. Black-chinned and Rufous Hummingbirds had also been recently spotted in the area, as well as Say's Phoebe, the Phainopepla, and Bullock's Oriole. However, the directions were poor, and we had little idea where any of the landmarks alluded to were situated. Apps like this are indispensable if you're birding without a guide, and when there are no other birdwatchers around. Birders will log on and share sightings of not just rare and endemic species but everyday birds, too, building a great picture of the local hot spots and the birds you're likeliest to see there.

We found ourselves on the outskirts of the town, at the edge of a small clearing, waiting around, staring at our phones as if they might magic up the birds.

We were about to give up and return to the car when we were approached by Russell, a local birder, who informed us he had some great birds in his garden, and would we like to see them? We most definitely would.

Hummingbirds milled around the feeders he had strung up on his porch, and I felt a stab of jealousy as I watched hundreds of these tiny birds diving from feeder to feeder. If you live in the right place, you don't have to wander the hills and valleys in search of the miraculous; you just have to put out some food and the miraculous will come to you.

We stayed and watched the hummingbirds, the Rufous and Black-chinned obliging us with stunning views, while Russell talked about growing up on the Paiute-Shoshone reservation on the eastern side of the Sierra Nevada.

At this point in my life, I wasn't aware of just how devastating an impact colonization of the Americas had had on its Indigenous peoples, forcing them into poverty and into yielding their lands. Russell had had a tough childhood: His reservation had been neglected and was subjected to regular water shortages. There had been little infrastructure for formal education. In his early teens, he had turned to the outdoors for solace. Hiking in the woods, mountains, and forests of Lone Pine fueled Russell's growing desire to work in the environmental sector. He broke out of the cycle of poverty, seeking education and bonding with nature.

As an Indigenous American, he had a significant connection to the land—it was part of his heritage—and, now a biologist, he was using nature to help others find their way back to that connection and local disaffected teens to find their own ambition.

In the years to come, I was to reflect often on this encounter— how, with his simple ambition, he had moved away from the reservation only to return in a way that was meaningful to his heritage and his career.

That evening, I thought about my own ambition: It was one thing to take VME people into nature, into the countryside, but the deeper issue was of engagement. Walking, hiking, and birdwatching made me feel good, but I had grown up with these pursuits; I felt better outside than in. If the countryside were to resonate with a wider, possibly semi-estranged audience, it had to deliver the same well-being to them. I decided I needed to reach further and do more. What I didn't know was that within just a few months, my encounter with Russell would lead me to action that would create fundamental change in the environmental sector.

Mum was mostly enjoying herself, but once again, she hadn't been great at booking accommodation. More than a couple of times, we arrived late in a town to find no vacancies in motels, hotels, or guesthouses. So we'd have to leave, find another town, and hope they had some space for us.

Mum was also, worryingly, edging toward mania, all part of the ebb and flow of her bipolar disorder. A couple of times, all perspective lost, she had egged Dad on to keep driving in search of one rare bird or another, so convinced was she that if we just pushed on a little farther, we'd find the bird of our dreams. And Dad did as he was told.

More than once we found ourselves on unsafe roads in the middle of nowhere because of Mum's hopeless directions. "If I'm driving all day, I want to know we have a meal and a bed at the end of it." Dad had said this exact sentence countless times during the trip, and he was right, but that was also true for Mum and for me. It wasn't her intention to strand us on the highways of America. For good or bad, this was our family's narrative on holiday, and probably would be as long as the three of us went traveling.

Santa Cruz Island, our final significant stop, is the largest of the Channel Islands off the coast of Southern California. With mountains, deep canyons, springs, streams, and over a hundred kilometers of dramatic coastline, it is home to our target bird, the endemic Island Scrub Jay.

From the boat we stepped onto perfect white sand, before a vista of rugged cliffs dipping into the sea. If there was ever a picture postcard of a stereotypical Californian idyll, this was it. The other tourists ambled off for a gentle walk around the island. We, of course, were on a mission.

As we made our way along the beach, following a line of stunted bushes, and up onto a coastal path, the Island Scrub Jay appeared— and not just one but three. As they perched up among the chaparral, their bright blue plumage glowed in the sunshine. I decided they were perfect for the island, their bellies the color of the sand and their azure-blue plumage an echo of the sea.

Our mission accomplished, we rejoined the other tourists ready for the boat ride back to the mainland. Time to relax? Of course not. We scanned the ocean with our binoculars, always looking for the next bird. Two Black-vented Shearwaters skimmed the tops of the waves, dipping and twisting on stiff wings, saluting us as they flew by.

Our trip to the US had exceeded my birding expectations. A combination of modern information technology and friendly, helpful local birders—the old and the new perfectly combined—had allowed me to see nearly two hundred new species for my world list.

Before I arrived, I had clear expectations of America. I was excited about yellow taxis, giant pizzas, diners, and iconic architecture, but in the weeks and months that followed our trip, those things were the hardest to recall. In the end, it was the outdoors that remained fixed in my memory: the giant redwood forests, the towering rock formations, the waterfalls, and the wide-open skies.

While I settled back into the rhythm of a new school year, I was still waiting to hear back from the organizations that attended my Race Equality in Nature conference. I felt that they had been given a short-cut to solving the issues at hand: all they had to do was act on them, and then tell me how the ideas that they claimed had so inspired them were taking shape.

All the waiting was for nothing, though. They had been given a blue-print for tackling the lack of diversity in nature, and while they had seemed enthusiastic, receptive, and grateful to the speakers, they hadn't run with the ball. Maybe attending the conference was all the effort they were going to make. The lack of any kind of response made me feel cynical and a little naïve. Had it been a massive waste of time, energy, and effort? I knew that if I didn't follow things up, then nothing would change.

Part of the problem, as I said earlier, was that none of the CEOs had come to the conference; instead they had sent their juniors. I needed to reach those who had the power to make genuine change happen if we were going to create a more diverse nature sector. There was no independent body to draw attention to the lack of diversity in nature; instead, I was a lone voice fighting to be heard through the deafening silence of inertia.

In September 2016, I set up Black2Nature, a charity whose sole objective is to increase the engagement of VME people in nature. Black-2Nature would be the official body and resource to help encourage the ambitions of the nature sector if they were serious about finding solutions. This was the platform I needed if I was going to force these organizations to become advocates of change.

Dad had suggested my first task would be to draw up a list of "core values." For example, how health and safety is a core value for all business models, which means the well-being of its employees is at the forefront of a company's vision, and therefore a universal consideration. Diversity would have to be a core value, an issue that must be addressed every single time a new scheme or project is launched, each time funding is to be allocated, every time a new reserve is opened. In this way, diversity could not be sidelined; it would be integral to the organization's values.

I had understood by this point that I needed to be more energetic and elbow my way into the corridors of power if necessary. Once more I wrote to the people in charge and in clear, firm, and *forceful* tones, I explained what *I* had done to move along the conversation and asked them what *they* were doing. I was at the beginning of something that would pivot me toward active campaigning, including, to my dismay, uncomfortable exchanges with reluctant CEOs. These overwhelmingly older White men did not enjoy being challenged about the racism within their organizations.

Campaigning is driven by frustration: it has its own momentum. The more I did, the more involved I became, the more compelled I felt to push harder, to add to the conversation. My activism felt instinctive and powerful. I began to attend and speak at an increasing number of forums in the nature sector about lack of diversity.

In 2016, I met the teachers at the Association for Science Education and Geographical Association conferences to talk about educating our future environmentalists. I discussed how VME students must be engaged on issues of climate change by making the issues relevant to them. For example, by talking to a Somali child about the drought and flooding in Somalia, and the prospects for the country ten years from now if no action is taken, the crisis becomes immediately relevant, urgent, and engaging, especially if that Somali child has relatives living through the crisis.

Around this time, I was growing increasingly aware of the echo chamber of the internet around climate change—it was very loud, filled with groups of people saying the same old things to each other. These people included celebrities, CEOs of nature organizations, and high-profile campaigners. Zooming out, I found that within this nature sector, only 0.6 percent of jobs are filled with VME people, and these include the organizations' cleaners and caretakers. The sector is predominantly White and predominantly male—it's easy now to understand the uniformity of opinion, the same voices saying the same things to one another, on social media, in the press. If our key influencers come from a homogeneous group repeating the same tropes, where is the diversity of thought? What do the people outside that homogeneous group think? Who is representing their ideas and their experiences?

In 2016, I took to social media to argue against a proposal to introduce a new Natural History GCSE (General Certificate of Secondary Education). On the surface, it sounds fantastic, doesn't it? What's not to like about a subject dedicated to exploring our natural environment, the wildlife and their habitats? But who is going to take up this GCSE? We already have three science subjects, one of which is compulsory. This is exactly the sort of subject that is offered up at private schools, which tend to have the resources to employ dedicated natural science teachers. It will also be a subject taken up by those already interested in nature. State schools, with their limited resources, offer only a limited range of GCSEs, and are unlikely candidates for a new one. This led me to one conclusion: the nature sector would thus become even more rarefied and elite if the select few able to take the exam decided to pursue a career in environmental studies. I suggested instead that the best, the preferred, and by far the most sensible option would be to introduce natural, climate, and environmental education throughout all subjects. For example, in English, we might be offered the chance to study nature writing; history and geography might give special attention to the ravages of climate change and climate activism through the ages; and so on.

I use this example to show how I was finding my own voice and that I wasn't scared of swimming against the tide. Campaigning, debating, and speaking from experience gave me the confidence to talk at the New Networks for Nature conference in 2017 to challenge the introduction of a Natural Science GCSE. The status quo was not going to stop me.

In October I shared a stage with environmental and political activist and writer George Monbiot and Green Party politician Caroline Lucas at the Festival of the Future City, where I argued that in order to have a sustainable city—a city in which people work together to minimize our detrimental impact on the environment—all inhabitants of that city must be engaged to work together. There is no solution if there is no majority consensus, and if VME people are not in the conversation in the first place, then we have little hope of achieving our environmental goals.

A vocal member of a minority ethnic group, I was nipping at the heels of the echo chamber, and I finally felt like my voice was starting to be heard.

{ 11 }

Here Be Dragons

WALLACE'S STANDARDWING

Wallace's Standardwing is a bird of paradise found only on the islands of Halmahera and Bacan in Indonesia. It was named for Alfred Russel Wallace, the first European to describe the bird. Wallace's biological studies of the islands of the Malay Archipelago led him to independently think up the concepts of natural selection and speciation at the same time Charles Darwin was developing his theory of evolution. The two men corresponded regularly before the publication of Darwin's On the Origin of Species.

The male Standardwing is polygamous, gathering in groups, or leks, to perform its spectacular aerial displays. To the accompaniment of loud screeching, they first fly upward before "parachuting" down, wings spread, through the branches of the dense rainforest trees. They flash iridescent green feather breastplates and shake their standards: long, white, pennant-like plumes attached to the wings. All this action is primarily geared toward asserting their superiority within the lek, but the watching females will choose to mate with the dominant male or males closest to the center of the display.

My heart in my mouth, eyes to the skies, I stepped gingerly onto the rickety bamboo tower that felt like it was hundreds of feet above a forest in West Bali National Park. (Obviously it wasn't; did I mention my fear of heights?) I was about to fulfill a childhood dream. The ground below was dry and dusty, the forest sparsely populated— all the better for spotting a Bali Starling, a bird I had first laid eyes on thousands of kilometers away, in captivity.

I can't remember the first time I visited Bristol Zoo, but it is as much a fixture of my childhood as riding a bike and bedtime. I insisted on spending a little time with each and every animal, so these were usually very long visits. Naturally, and over the years, my interest narrowed to the bird exhibits. I had been obsessed, at five years old, with the Rainbow Lorikeets. Armed with a tiny container of sugar water, I would enter their enclosure, and the many-colored parrots would perch on my fingers as they dipped their heads to drink.

But it was the Bali Starlings that kept me coming back. By the time I could read the information plate outside their enclosure, I had already fallen in love with these glowing white birds. With black wing and tail tips and a crest of white feathers, they have a certain regal quality to them—to me they looked like judges about to preside over some somber court case, their heads cocked, ready to point a wing and declare, *Guilty!*

But the species was in trouble, surviving only in very small numbers on the island of Bali, and the province's only endemic bird. Illegal trapping, and a worldwide thirst for caged birds, had driven its decline.

And any minute now, I was going to see them in the wild. A dream come true, you might think. I thought so, too, until I had to climb the bamboo steps to a platform many meters above the ground.

There was nothing to do on this Indonesian bird tower but look for birds. The sun was hot, the forest shade below cool and tempting, but I had no intention of moving while there was still a chance of seeing the starlings.

Waiting, watching, and positioning and repositioning the telescope to scan the treetops, we eventually saw a trio of blurry birds through a heat haze on the horizon. They could be Bali Starlings or they could be something else—like the almost equally rare Black-winged Starling; it was difficult to tell. Rare birds are a big deal: you have to be 100 percent certain of your ID. The heat haze wasn't helping either, so we waited some more. The birds took flight and pivoted in the sky, joined by nine others, and now they were flying directly toward us.

Our guide, eyes glued to his binoculars, suddenly let out a yelp. "Here they come!"

I had the perfect view across the forest, across the sandy bay in the distance, and out to sea. I held still—my breathing was steady as I waited for the Bali Starlings to approach. The forest disappeared, the sea was gone; there were just these tiny, brilliant birds and me.

In 2005, there were fewer than ten such marvels in the whole park, and that day, in 2017, I was staring at twelve. The stark contrast between the captive birds in the zoo and the dignified creatures streaming overhead made me hold my breath until they were out of sight. They were exactly where they belonged: in the sky.

It was Mum who had really wanted to go to Indonesia for our next trip, having never been. Dad had already spent a significant amount of time birdwatching on this nation of islands, but there was still plenty more for him to see. The sheer number of endemics and rare birds was astonishing, and for Mum and me, they would all be "firsts." It would take years to explore each of Indonesia's islands properly, but we were hopeful that six weeks was long enough to land us our key targets.

My family is unusual in that we mostly prefer solo birding missions, organizing our own transport, accommodation, and local guides. In general, however, world birders tend to tour in groups. Mum and Dad have always enjoyed the freedom to go off the beaten path without the rigors of an unyielding schedule to get in their way, and there had been good reasons why we traveled alone. When I was very young, my parents worried that a full-on guided tour might be too much for me. The other concern was Mum, of course: she loved the chaotic way we stumbled around the world, the fact that we did *everything* together, and Dad hadn't wanted her to feel restricted by a timetable or the whims of other birders. But I wasn't a kid anymore, and Mum seemed okay with the idea of traveling in a group.

However much we loved traveling as a trio, perhaps now was the time to spread our wings and try birding with others. We organized our trip around key target species and then sought out like-minded birders to join us, a maximum of three so as not to outnumber our triumvirate.

Meanwhile, as the weeks sped by and the date of departure approached, Mum's moods, once again, were seesawing. Introducing third parties into what is already a stressful situation began to feel risky. Traveling with others, living in close quarters, birdwatching together—it can be intense at times, with little space to deflect any frustration. Was she going to do or say something "crazy" or, worse (I was fifteen), *embarrassing*? Might she completely misjudge a social situation, an innocent exchange? How would she react to Dad and me if we tried to steer her behavior? The classic "Why are you kicking me under the table?" query from her as we try to divert her from one subject to something less contentious is all too familiar in my family.

We had some idea by now of the things that tend to trigger Mum's temper: missing birds, skipping her meds, losing stuff, and lack of sleep. There was obviously plenty of opportunity for all these things to happen on the trip, but there was no way to know how other people might react to being on the receiving end of an outburst.

Before we had time to reconsider, eager replies had come in, and the extra people were signed up. Mum, it should be noted, didn't share any of Dad's reservations; she was looking forward to the company.

The truth is that Mum *loves* being away. She loves traveling, the car journeys, the treks up the sides of mountains to get her bird, eating together, moving from place to place, the bustle of the chase, and the thrill of the twitch. Deep down, I knew that as long as she carried on feeling like this, while there might be a few hiccups along the way, she wasn't going to let anything ruin her adventure, or mine, or Dad's.

And so far, the trip *had* been a success. We traveled with three fellow birdwatchers whom we had bumped into a number of times over the years at various twitches in the UK. They were pretty hard-core, asleep by sunset and up before dawn, which suited us well. The most annoying thing about some other birders is dawdling—waiting for others to catch up is intensely frustrating. But fellow birders can do other irritating things too! Talking too loudly when your target bird eventually shows up, celebrating seeing the target bird when you haven't spotted it yet, and so on. It's a long list, but only minor annoyances when you consider the huge positives: the camaraderie, the helpfulness within the community, the mutual joy when a rare bird appears, and the delicious knowledge that everyone is feeling the same way.

Fortunately, our companions were as committed as we were, while avoiding the bad habits outlined above. For the first time, I felt as though we were birding with people who were as intense as my own family.

Our trip to Indonesia had begun on the island of Sulawesi, in Lore Lindu National Park, where our target birds were specifically the endemics, such as the Sulawesi Hornbill, the Piping Crow, and the

White-necked Myna. They came to us easily enough, whether we were deep in the forest or on its edges.

On our first morning, we were up at three to begin the drive to the park. It was still dark when we entered the forest, using torches for the first hour of the hike, trying to identify the multitudes of birds belting out their dawn chorus. The sun was rising by the time we emerged from the trees at the top of a steep hill to begin walking the mountain path.

The park is a sprawling landscape of lowland and mountain forests, rivers, lakes, and even megaliths. As we were walking along the path in the growing heat, our target bird was the Satanic Nightjar, a bird only officially rediscovered in 1996 following the collection of a single female specimen in 1931.

It didn't feel as if we were in Indonesia, with its popular images of white sandy beaches, pristine clear-blue seas, and lush rainforest. We were stomping along on an unphotogenic, rocky mountain track.

Nightjars in general, whether in the UK or on an island in Southeast Asia, prefer to sleep during the day, *on the ground*. While they're not a rare family, their plumage of brown hues renders them almost invisible in their habitat. They are also very small. We hiked into the afternoon, noting other birds along the way, including a Scaly-breasted Kingfisher, a Purple-bearded Bee-eater, and much later, a Hylocitrea, endemic to Sulawesi and only ever seen on the mountain trail. Eventually, in a rocky clearing in some random slice of woodland, and after much bush prodding, we were rewarded with a pair of Satanic Nightjars comfortably installed in the undergrowth—waiting for us. We wouldn't have seen them had they been asleep, but their eyes, wide-open globes of liquid black, marked them out.

"They're really cute," I noted. "What's so satanic about them?" No red eyes, no horns or smell of sulfur, just small, squat birds.

"Their call sounds like the ripping out of a human eye, apparently," said Dad helpfully.

At that moment, one nightjar opened its beak, revealing a wide pink expanse, and let loose its lilting, pirruping song; it was a sweet sound, nothing gory about it at all.

We were a week into the trip when I began to feel a little uneasy in our group. Before we came away, I hadn't been on a twitch for weeks; with GCSEs on the horizon, I had been studying hard, while also setting up Black2Nature. Arriving in Indonesia, meeting the others, and bird-watching once more felt like the most natural and spontaneous thing I had done in months. And yet, I was suffering from a mild case of im-postor syndrome. Given I had just achieved the four-thousand-species landmark, I should have felt anything but paranoid, but my confidence hit a low among this group of very serious birders. I was just a young VME girl, an anomaly in a mainly White, mainly male arena, and I felt, for the first time, a little out of my depth.

By taking on the persona of Birdgirl, I felt I had somehow declared I was an absolute expert when it came to birds, the fount of all bird knowledge, but that had never been my intention: I simply loved bird-watching. Inadvertently, I had shined a spotlight on my hobby and opened myself up to scrutiny on social media and in the wider birding community. I wasn't competing with anyone's list but my own—and maybe Dad's, sometimes.

Given that the birding community is dominated by older men, I had always had an underlying suspicion that I was in the wrong crowd. Birdwatching was becoming my refuge as I got older, a space in which I could lose myself to the skies, whether I noted down my twitches or not. I really didn't want the competitive aspect to encroach upon this space.

Was I a fake? Some days I felt as if I had woven a complex deception online, however unintentionally. And now, paranoia had set in, making me worry that at any moment, one of our group might announce to the others that he would no longer be fooled by this young girl, enough was enough!

When birds move, they're easy to pick out—it's when they're stationary that they're harder to see. I have great eyesight and was better than everyone else at spotting immobile birds *and* giving directions to their exact location. Over the following days, I recognized that these two "skills" somehow endeared me to the birders in our group. Very slowly, my confidence returned. I realized no one gave two hoots if I was a good birder or a bad birder—they just didn't want me or anyone else to get in their way, and this suited me because I felt the same. It took a little while to shrug off village life and the constant studying, but in a few days, I was back in my comfort zone, binoculars pinned to my eyes, capturing new and rare birds—impostor syndrome vanquished.

One of the most beautiful birds of the whole trip was Wallace's Standardwing, a bird of paradise endemic to Indonesia. We would have to go to the volcanic island of Halmahera if we were to have any chance of spotting it. Birds of paradise are known for the spectacular plumage worn by the males of the species, and I was keen to see at least one bird flex its feathers in a mating dance or in its own display ground, or lek.

We were once again in mountain rainforest. Even though dawn was yet to break, it was humid, and a thick mist hung in the air. We had made our way down a slippery forest trail, our headlamps picking out obstacles ahead until we reached the clearing. To our untrained eyes, this space didn't look so different from the rest of the forest, even though, to the standardwing, it was a theater awaiting its finest performance, and we were its audience, standing by in silent expectation for dawn to light up the stage.

The Wallace's Standardwing is named in honor of the British naturalist Alfred Russel Wallace, the first European to describe the bird, in 1858. It was from the island of Halmahera that he famously wrote to Charles Darwin with his thoughts on evolution, which Darwin went on to consult in support of his own theory of natural selection.

Wallace had brought back the body of a male standardwing to the UK for further observation, but never saw the bird perform its mating ritual. I imagined the somber-suited scientists handling the jointed limbs and various plumed appendages of the bird, all the while scratching their heads and wondering what purpose these curious features had. If the explorer had just waited for the bird to dance before he shot it, all would have been revealed.

I wanted to see the Standardwing, but most of all, I was dying to see his moves.

We crouched behind the stubby bushes, stock-still, our eyes trained on the dusty clearing. It is true to say the standardwing is not as brightly colored as other members of its family: it is a small bird, the size of a thrush, and it is also mostly brown. This much I knew, so when it hopped into its lek, I was surprised. No photo I had ever seen had shown the true beauty of this bird.

We were silent, barely daring to breathe, as the standardwing strutted the circumference of his stage. A bright orange protrusion above his bill caught the rising sun and glowed. What looked like an extra pair of bright green wings extended from his shoulders, and then something remarkable happened: long white plumes fanned from his wings, mothlike, as he started to dance. The bird began to call, its cries loud, echoing into the trees. He seemed to quadruple in size in front of my eyes as he puffed out his chest, iridescent emerald "wings" shooting out to catch the sun. *Choose me!* he seemed to be saying as he parachuted between branches. And how could anyone refuse him? This was a bird of pure spectacle, never mind paradise. I was immediately back in the documentary *Attenborough's Paradise Birds*, where an equally enthralling creature interrupted Sir David time and again with its dance, its song, and its insistence that all eyes should be focused on *his* display, rather than on the guy who wouldn't stop talking.

As we walked back, display over, faces flushed with success, Dad burst my happy bubble by asking how much longer I could see myself coming away with them. Maybe I would prefer to go away with my own friends? I'd be sixteen soon, then seventeen....

It was the first time I considered not spending my summers with Mum and Dad. What would that even look like? I imagined going on birding trips with my friends, but...which friends? Did this mean no more birding trips? Or would I have to go on my own? "I don't know," I eventually replied. "Do I have to decide now?"

Dad laughed and pulled me in for a hug. "You can come away with us until you no longer fancy hanging out with your parents."

He has checked in with me over the years since; it's his way of making sure I still *want* to take trips with them, to show that I have a choice. But on that volcanic island, I realized I couldn't necessarily do this forever. I would be taking my GCSEs and then my A levels, and afterward I would hopefully go to university. Eighteen-, nineteen-, and twenty-year-olds don't spend their entire summers with their parents, do they?

As usual, we set out before dawn. Ten minutes into our drive, Mum realized she had left her binoculars behind. The driver turned the car around without a word, already used to these lapses.

Our target bird for that morning was the Ivory-breasted Pitta, endemic to the island of Halmahera and its satellites, yet here we were retracing our steps in tense silence, moving away from the twitch, not toward it. It had become a familiar scenario over the years: Mum leaving some piece of kit either in a restaurant or back at the hotel or lodge. It wasn't the first time on this trip either. Mum as usual took her frustration out on Dad, accusing him of failing to remind her. Our three fellow birders were getting used to Mum's forgetfulness by now, and while she and Dad exchanged a few sharp words, they didn't have a full-on row, which was a relief. I decided having others close by was helpful.

Binoculars recovered, we turned around again and headed toward the edge of the woods. From there we walked along the side of the road that bisected the vast forest, peering into the undergrowth for the skulky Ivory-breasted Pitta. Unlike many elusive birds, this pitta is beautiful—well worth the hours you might have to spend waiting to see one.

It's always surprising to non-birders that roadside birding yields such riches, but in remote locations, this is the best place for spotting the maximum number of birds. Clear paths into dense woodland in tropical countries are rare, and the visibility once inside is low, so we tend to gather on the edges, even as traffic roars by.

Calling the pitta while horns blasted and the tarmac boiled didn't make for the most enjoyable birdwatching, and the pitta's song was frequently drowned out by the passing traffic. "We have to go in," Dad suggested finally, à la Indiana Jones. "I can't hear the recording, never mind the pitta."

The six of us followed our guide into the forest, where it was instantly much darker and more humid; the heavy canopy blocked out the morning sun but little of its heat. Still no sign of the pitta. By this point, we were all feeling a little desperate, but it is also true to say that the most exciting birds—because of their beauty, rarity, and extraordinary behavior—are often the hardest to see.

Our guide began to whistle for the pitta. Trilling notes poured from his mouth, better than any recording. Calling and waiting, calling and waiting, until finally, the Ivory-breasted Pitta hopped into view on long legs. Despite the gloom of the forest, the bird stood out like a beacon. Holding my breath, I stared, drinking in the marvel, the silvery-blue and green wing patches contrasting with its black upper side, and there, burning like a hot coal on its belly, was the red flame licking the white of its breast. And then, seeing us seeing it, it hopped back into the bushes. It never ceases to amaze me how a bird as brightly colored as the Ivory-breasted Pitta can just melt away, vanishing into the undergrowth.

Wide grins and hugs all around, I forgot the heat of the day, the cloying humidity, and felt reenergized by this little bird. Whatever size it was, and however briefly it stood in our midst, its presence was huge and lasting. The Chew Valley might as well have been another planet in those moments; my real life felt so far away, like a dream I could barely recall.

Occasionally, during episodes like this—when we've had to work hard for our bird, walking, waiting, watching, and calling—my family begins to operate like a simple organism, each of us in sync with the others, as we make eye contact; nod directions to follow a lead, a path; or glance into the trees, intuitively understanding what we have to do. The chase makes us a team. Like all families, we argue like mad at times, but there is nothing like seeing a special bird, one that calls for extra effort, that makes it all worthwhile. Mum's fury and Dad's irritation about the neglected binoculars dissolved. They forgot themselves, we all did, swept up by the challenge of spotting a rare and extremely gorgeous bird. In the same hopeful footsteps of the pitta, we returned to the car, the Craig Family Harmony Index restored to balance.

In my experience, other animals rarely outshine birds, but even I had to admit that the Komodo dragon was one such creature. On the last day of the trip, we went to the island of Komodo. At dawn, the huge lizards are cold and immobile; later, when the sun starts to pound, they move fast and eat everything, including smaller Komodo dragons. It was early in the morning, which is when it is still safe to wander among them without fear of being chased. In the heat of the day, these mighty lizards grab a spot on the beach to bask in the sun and refuel their energy stores, giving them enough energy to zip across the sand, snapping at the heels of tourists who have ignored instructions to avoid the beach. The potent cocktail of venom in their mouths weakens their prey when it falls victim to their bite; the Komodo dragon then wanders off, seeking out other prey, only to return when their quarry is too immobilized to fight back or run away.

The island was a sandy expanse of twisted trees and dead grass, an apocalyptic landscape, baking on the edges of a luminescent green sea. The only living things seemed to be the massive lizards. Our guide carried a long, forked stick, which he would have to wield if a dragon woke up early. We trod carefully. They were everywhere, slumbering peacefully, and it was hard to imagine them any other way—until we left. The heat was up when we climbed back into the boat, and the dragons, pronged tongues flicking in and out, lined the beach like holidaymakers about to begin a day of sunbathing. More dinosaur than gecko, they were terrifying and thrilling, and were now beginning to swarm the island in search of a meal. It was time to leave the island and Indonesia.

The Harmony Index was at its peak when we left. We had had the sort of trip we had been trying to achieve for years. Mum's moods were mostly even, Dad was rested, and the birds were spectacular. One of the reasons for this was that we had shared the trip with others. It helped having people around, we realized, giving Mum other outlets and distractions, which in turn allowed Dad and me to relax into the adventure. Mum is usually a tyrant about birding, keen to see everything on her list and more—time off is forbidden. But in Bali, she even let us go scuba diving!

Mum had enjoyed the trip, but back at home, the first signs of mania were surfacing. She began to stay up late, unable to tear herself away from the computer, buried in research about whatever had taken her fancy that day, week, or month. Dad wasn't faring much better. He couldn't settle down until Mum came to bed. He was not only exhausted but also frustrated that Mum wouldn't listen to him. Whatever Mum was up to was more important than any of us. When she did eventually go to bed, she would fall asleep instantly, but Dad would lie there, desperate to nod off but too wound up to sleep.

As the weeks passed and winter approached, Mum was growing ever more belligerent and oblivious to Dad's entreaties that she see the doctor. It was the first time I felt seriously worried about *his* mental health. We were this traveling trio, this "us against the world" gang. However tough times had been, we had challenged our issues together. We had used our trips and birdwatching as coping mechanisms, and now Dad had planted the seed that they might be over. How would we cope if we didn't travel as a family anymore?

In some ways, I was the glue that held us together. As long as one of them worried about the effect Mum's illness would have on me, there was more impetus for them to address it. If I weren't there, what would happen to Mum? How would Dad cope?

I love my mum deeply, but it is tiring and frustrating living with someone who has bipolar disorder. It's always changing, and just when you think something is sorted, the next curveball will be hurled in your direction. It's easy to understand why her depressive periods are so upsetting: she is distracted and becomes disconnected from the family. This is stressful because we don't know what she's thinking about. But time and again, it was the manic episodes I found infuriating, especially when I was younger, when she became obsessive, staying up late on her computer, delving ever deeper into whatever it was that sparked her interest. Or she would throw herself into a neighbor's problems, or a cousin's relationship issues, devoting all her time to them. It felt like an active rejection of me in favor of someone or something else.

Every row was followed by an explicit conversation in which Dad or I would point out to her how her behavior made us feel. But things rarely changed when she was in the grip of mania.

Omnipotent Helena believes she can fix anything. Whatever her current focus or obsession is, she throws herself into research, losing time to tunnel vision. These issues are generally around conservation and diversity, and Mum knows how to drill down. And just as quickly as it appeared, she drops one obsession and moves on to the next.

But there is a flipside to Mum's mania: things tend to happen because she makes them happen—Black2Nature, for example. In my work, Mum's drive has been invaluable in achieving a platform.

These days I understand there is a gap between her understanding and acknowledgment of her actions and her capacity to do anything about it. I can rationalize it, I could even rationalize it when I was younger, but it was and still is upsetting because it *feels* willful.

Around this time, Dad mentioned he felt like he was constantly trying to plug holes in a leaky dam—he dealt with one thing and another popped up. By nature, he's organized, the kind of person who likes to make plans and stick to them; it's how he copes. But he is also sensitive and often hurt by Mum's gibes. It's difficult not to take things personally even if you can rationalize moments of confrontation as a symptom of Mum's illness. Her words can be wounding, and Dad's coping mechanisms, and mine, were being stretched.

The end of that year was marked by my first panic attack—or rather it was the first time I recognized that what was happening to me was a panic attack. In the past, I had had episodes of anxiety, usually before a media interview, or even during something as simple as a row between my parents.

Toward the end of 2017, a couple of months after the best holiday of our lives, I watched my parents screaming at each other from the top of the stairs. Dad was insisting Mum's meds needed adjusting, that she had to go to the GP first thing, that he was at the end of his tether. But Mum was having none of it. She was fine, better than fine, in fact. Hadn't she just had the time of her life in Indonesia? Why couldn't he be glad, just for once, that she was happy?

A new fear gripped me. Mum's eyes were shining, she was bristling with energy, nothing about her suggested she was feeling suicidal, but Dad looked rough, worn out, and sad. He grabbed his coat and slammed out of the house into the dark November night. I would have cried if I could have caught my breath, but it was strangled up in my throat. My chest started to heave, and I collapsed on the top step, my head on my knees, shaking all over. Mum was up the stairs in seconds, her arms tight around my shoulders, telling me to take long, slow breaths. We sat there for fifteen minutes while I tried to regain control of my lungs, Mum all the while soothing me with words she would have done well to heed herself.

Dad, despite the stress of caring for her, has always insisted he's not going anywhere, that he would never leave us. And Mum never once thought he would walk out on the family. While I believed him, I didn't know that he wouldn't snap under the pressure of being Mum's carer and have some sort of mental collapse himself.

"You need to be kinder to Dad," I was crying, hiccuping. "You need to listen to him!"

"I will, Mya, but you have to calm down."

"What's going to happen to you when I'm not here?"

She didn't have a response to this, which, given she was manic, was unusual. She just carried on holding me, rocking me back and forth while I talked about Dad, how he and I had managed things together, how he needed my help. How would he cope if I wasn't around to support him? We made a complicated puzzle, the three of us; with a piece missing, the picture would be ruined.

"We'll be okay," Mum said. "And none of this is your responsibility."

My panic attack was a bit of a wake-up call for Mum. It had lasted fifteen minutes in total, but to me it felt like thirty seconds. I might even have blacked out for some of it.

When Dad came home, Mum told him she would visit the GP the next morning.

Her antipsychotics were increased, and Mum's mania receded. She also noted that while that evening had been one of the worst we had experienced as a family, she had gained some perspective. She would try to be more alert to how her behavior was affecting the family, and take action before things went too far. Seeing me so upset had unsettled her; both Dad and I were hopeful that she would get better at recognizing the signs that things were heading south, but this was a promise she had made in the past so many times. My mum is sick, and there is only so much she can reflect on or feel objective enough about in order to to change her reactions, her thoughts, her convictions. This is why it's crucial to get her meds right, and find the perfect combination to give her a little more distance from her illness, and give her the space in which she might look at some of the things that hurt not only her but all of us.

The Eighth Continent

HELMET VANGA

The Helmet Vanga, with its enormous, arched blue bill, is perhaps the most iconic member of the group of Madagascan vangas. Found only in undisturbed humid rainforests in the lowlands of northeast Madagascar, it faces the very real possibility of its ecological niche completely disappearing by 2050 because of the dual threats of habitat destruction and climate change. With no ability to survive outside these forests, the Helmet Vanga may well be driven to extinction. In 2050, I will be forty-eight years old.

In 2018, I got my GCSE results in the scorching heat of a dry Madagascan desert.

"No way! *No fucking way!*" I screamed into my phone. Back in the

Chew Valley, Ayesha was at my school, and having finally managed to get through, she was reading aloud my results, the voices of happy teenagers screeching in the background. I had been nervous; it was the first year we were to be graded using a 1–9 system, instead of the usual ABCDE. My parents were leaning in to hear what my sister was saying. Even our guide seemed interested, but maybe it was the sight of me jumping around and cursing that had piqued his curiosity. I had been all business up to this point, following him dutifully across a barren, sandy landscape. But my results were good, despite the challenges at home, and easily worth a curse or two.

Driving from the capital city of Antananarivo toward the far north of Madagascar, we had been desperate for a bar or two of connectivity on our phones. We were already forty-five minutes late for the appointed exchange with Ayesha. The moment a signal appeared along this desolate road, we jumped out of the car.

We were on our way to Ankarafantsika National Park, nine weeks into a ten-week trip and just days away from returning home. This was the first time I had thought about school since zipping up my pencil case at the end of my final exam. It was just as I was thanking Ayesha for making my day that our guide turned away from our huddle, his arms shooting into the air, index fingers to the sky.

"Malagasy Harrier!" he yelled.

Dad and Mum and I snapped our binoculars to our eyes. Exams instantly forgotten, I focused and refocused the lens, desperate to confirm that the bird we were looking at was indeed the Malagasy Harrier.

"I can't believe it," gasped Dad, shaking his head.

"This has to be one of the best days of my life!" Mum said. I gave her a look—just moments ago, she had been hugging the life out of me, so proud of her diligent daughter. But honestly, what did exam results matter to any of us now, in the presence of the mighty harrier?

The Malagasy Harrier is an endangered and extremely rare bird of prey; it's in the "once in a lifetime" category. A pure embodiment of strength, it sliced long, slender wings through the air, but there was also a delicate grace in the way it turned in the wind. As though it had sensed our rapture, it continued to circle above our heads, graciously giving us its time and a display. There wasn't another soul around, and I felt a sudden tug of grief for all the birdwatchers who weren't here to enjoy the sheer spectacle of this bird as its pale underbelly, a streak of white against the deep blue of the sky, passed overhead. It turned again and, with a final wave of its wing, headed away from us and into the distance, dropping down once to the ground—no doubt to pick up some unlucky lizard—before soaring up and away.

"It came to celebrate your results," said Dad as the harrier faded away. "But if you'd got all nines, we'd probably have seen its mate too."

"All nines," he'd said with a grin, but he wasn't being serious; my parents aren't tiger mums. They want me to work hard, but they also know there are more important things than studying—like Malagasy Harriers. Still, the bird *did* feel like a reward, as my sixteenth year had been tough.

I had imagined the period directly after my exams as an exhalation of sorts, a sudden unwinding of the tension that had held me tightly coiled for a year. Studying hard, fitting in twitches, continuing my campaigning for an inclusive nature sector: all these required a level of commitment and organization that was hard for me to sustain. By July, with ten weeks in East Africa on the horizon, I was desperate to take some time for myself.

A weight had certainly been lifted from my shoulders with the exams out of the way, but deep down, so much more remained unresolved. The familiar longing to step outside my life and into a landscape so unfamiliar that it might as well be another planet had been irresistible.

Adolescence had only made me more self-conscious: like many

teenagers, I often felt awkward around others, at a loss for what to say or how to *be* in social situations. I found it less scary to talk to thousands of people online about systemic racism than to challenge the boy I sat next to in math about his Islamophobia.

I was busy, too busy. My blog was thriving: I had had over four million views. I was moving toward fifteen thousand Twitter followers, and I was reaching more people than ever to talk about birds and, increasingly, the lack of diversity among nature enthusiasts and the institutions that catered to them. Newspapers and nature programs alike wanted to talk to the "teenage Bangladeshi" about birds and Black2Nature. School was one thing, and everything else was another. I began to feel like I had a full-time job beyond school.

I felt like no one understood how much I was doing, and at the same time, I was somehow powerless to say no to any request that would let me talk about birdwatching, conservation, or anti-racism. I *wanted* to go on *Countryfile* and discuss my latest twitch, and to write a column for the *Chew Valley Gazette*—and if I didn't have the time, I would forgo hanging out with my friends, sleepovers, and even my regular hikes into the hills.

Black2Nature was gathering momentum, expanding to fill what little time was left. We were running more events to spread our message, and focusing on fundraising so we could carry on running events. I was invited to conservation conferences to talk about our aims, to give advice to the sector about their existing diversity initiatives and how they might implement others. I was usually a bundle of nerves before these speeches, chewing over the same question I asked myself before any such event: What could a sixteen-year-old tell these professionals that they didn't already know? But then, when I went onstage, my anxieties would fall away. I did have something to say and it was important, so they had better listen. But the seesawing of my emotions—the anxiety and doubts, the bursts of confidence and then more doubts—was eating away at me.

Adolescence is tough, and I was subject to its narcissism, choosing to believe that everyone had an opinion—and when it came to me, a negative opinion. By year eleven at school, I had doublethink down to a fine art: if I wasn't talking about my media work to my friends, then no one knew about it. I could pretend I was just like everyone else, doing the same things, going to the same after-school clubs, and socializing on the weekends. I was like a kid who thinks she's invisible just because she covers her eyes during a game of hide-and-seek. But the anxiety was always there, hovering beneath the surface, and I would brace myself for the moment my invisibility cloak was ripped away.

But what would it reveal?

Very simply, I was terrified of being judged. I had made some lazy assumptions about the other kids at school, believing the city kids among my schoolmates were more likely to mock me for birdwatching, blogging, and hanging out with Chris Packham than my countryside peers. They were altogether "cooler," more switched on to the nuances of adolescence and, I believed, quick to judge those they couldn't categorize.

It was after a national newspaper article appeared, in which I discussed getting trolled by Islamophobes on Twitter, that I was proved not exactly wrong but certainly *less right* about my friends. I was a little subdued in school the day after that article, wary that someone might have read it and would want to talk about it—that wasn't the kind of spotlight I was after. I needn't have worried, because it didn't seem as though anyone had seen it. Except... someone had. A girl who lived in Bristol—older and much cooler than me—approached the lunch queue in which I was waiting.

"I saw that piece about you. I can't believe anyone could be so mean to a kid," she said. "I'm so sorry you're going through it. You really don't deserve it. You're doing great work, you know, the stuff you talk about."

I barely heard her words. It was the fact that this "mean city kid" was talking to me about something I had said in a newspaper, issues I had blogged about. The hierarchy within school can be brutal; it was true there was a savviness among the Bristolians that often left the rest of us feeling a little rustic, but maybe that was more to do with us than any real hostility. I had felt that if anyone were to tease me for being "out there," it would be an urban kid, yet, right now, the opposite seemed to be true: none of my local friends had offered any sympathy.

"Thanks," I mumbled, stumped. And then I said it again and meant it: "Thanks."

She smiled and gave me a soft punch on the arm, and then it dawned on me. She was another teenager, just like me. Twitter, trolling, social media—*this* was our common ground. It wasn't about which part of the county we lived in; our language was mutual, instinctive in many ways. She understood online bullying—we *all* did. She was going through her stuff, and had enough empathy to show me some solidarity while I went through mine. The line I was drawing between my worlds had sheltered me, that was true, but it had also excluded me from instances of kindness, just like this.

This was the first time anyone at school had been positive about my work. The very fact that a schoolmate was praising my ideas, my public persona—the very persona that made me so self-conscious that I barely mentioned what I was doing to a single soul outside my family—gave me pause.

While it wasn't exactly a light bulb moment, I felt a little less lonely afterward, less judged, and more understanding of the complex makeup of the other kids in the school. Our adolescence unified more than it divided us.

. . .

The Islamophobia in school was harder to negotiate. There were plenty of racist kids, but most of them didn't target me in the same way they might discriminate against Black students or the girls in headscarves. I didn't wear a head covering, and so didn't overtly fit into the racists' category of "other" in the way other Asian Muslims might.

But I had been aware of Islamophobic attitudes from year eight. Ever since the American journalist James Foley had been beheaded by ISIS in Syria, some of the boys, taken by the idea of "gunning down Muslims," thought nothing of bellowing these sentiments in the classroom. I had never said a word. I felt that if I started an argument, I probably wouldn't win. The anti-Islamic sentiments felt too close to home and far harder for me to challenge than other social and environmental issues.

By year eleven, the racism had ramped up outside the classroom; it was now less macho bravado and more a specific intent to spread hate. The boys who had wanted to gun down Muslims still shared their crisps with me in year eight, but we were older now, and I had worked out that, as a Muslim, I was a member of their target group, even though I was never directly targeted.

On a random group chat on Snapchat with people from school, a few of the lads began to boast about their far-right allegiances. Essentially, they were idiots rather than active racists, but they were happy enough to share photos of themselves with pillowcases on their heads while making jokes about nooses and posting terrible drawings of burning crucifixes. Shortly after this, anti-Muslim and anti-Black jokes began to fill the chat, until eventually, I, along with a couple of friends, reported them to our head teacher.

The school did not respond directly, but invited the three of us to give a presentation to the teachers on why it is wrong to hate people of color, and how it sucks to go to this school if you're not White. We weren't saying anything revelatory—God knows the racist kids were well aware they had crossed the line, and the teachers also knew where this line was drawn. This was 2018, by the way, not the 1960s.

I had to question just how much had really changed since my grandfather's day. These days, Black people (and the Irish) can work on the buses, spend the night in a B&B, and drink in any pub in the land... but at the same time, sharing a photo of yourself in a white hoodie while espousing the language of White nationalist bigots was also fine, deserving of no greater rebuke than a lecture by three students, who were more than a little scared of a possible backlash.

After the presentation, our teachers were instructed to intercept racial slurs with a metaphorical slap on the wrist. This was the most they would ever do, an ineffective system of punishment that, for some reason, they felt compelled to update me on whenever they had caught a racist live in the act of disseminating hate.

In social studies, we were educated on racist language. "Is *Paki shop* an acceptable term?" we were asked. Some of the class believed it was perfectly acceptable. Weren't the owners of the corner shop from Pakistan, after all? And aren't White people who were born in the UK referred to as Brits? This was the level of rationale. Our teacher agreed, arguing that *Chink shop* was also acceptable, as it was used by older generations whose intention was merely to differentiate said shop from other shops in the high street. No racism intended. Meanwhile, and ironically, our school had just received an award for its pioneering work against homophobia and *for* LGBTQ rights, so they were aware of discrimination but felt absolutely no compunction to apply the same rigor to racism.

I exited the Snapchat group chat soon after the Ku Klux Klan had arrived; they were easy enough to avoid, but my Twitter feed was harder to ignore. Tweets featuring my latest twitch never elicited anything but genuine enthusiasm, excitement, and a shared love of birds.

It was only my messages about diversity in nature pastimes (and the lack thereof) or climate change that seemed to lure in the hordes of keyboard warriors who lurk in the dark corners of the internet, keen to tell a teenager she hasn't a clue what she's on about, and in any case, she should refrain from commenting on a country that isn't her own. Those spiteful missives hurt, although I didn't know the abusers and was unlikely ever to meet them, and their comments were juvenile at best. Why a tweet highlighting the rising temperature of the planet should make them so angry was genuinely baffling. I felt that old urge to shut the whole thing down. I swung between a reluctance to incite further abuse and emphatically deciding I'd say whatever the hell I wanted.

As my GCSEs drew closer, I withdrew from social media and lost myself in test prep. Weirdly, it felt like an escape.

After my exams, I decided that I would extend the break and forget about Twitter, school, and the Ku Klux Klan and devote every waking moment to birdwatching. In the autumn I would return to fight the fight, but right now, the first stop of our African holiday—Tanzania—beckoned.

Our trip to East Africa was filled with target birds, each one more intriguing than the last. I felt as though I were on a treasure hunt—the rules not so much about clue busting but about how to wait, watch, and listen—and the prize was so much greater (to this sixteen-year-old, anyway) than gold. Weren't the colors of the Green-headed Oriole more resplendent than cold, shiny metal? Adorned with a bright yellow plumage, this bird was liquid sunshine, its red eyes and beak glowing more brilliantly than any ruby.

And the Usambara Double-collared Sunbird, the males of which are a riot of red and green feathers—a squat bird with a long, curved bill, it had actually made me gasp while it hovered in the branches of a tropical tree in a lush rainforest. Whatever treasures were in the chest, this bird wore them all.

In Tanzania, we began birdwatching in the mountain forests of Usambara National Park. Because it held many such wonders, we had prolonged our visit, desperate to catch at least a glimpse of the Usambara Eagle-owl, endemic to the mountain landscape.

In our first lodge, late at night, we sat around a fire with some villagers. Mum, especially conscious of poverty in the countryside, had packed a vast amount of clothes to give away as we traveled. But there was one item that had yet to find a home: a brand-new Liverpool Football Club T-shirt. So far we hadn't met a single Liverpool fan. Until tonight, when, by the flames of a roaring campfire, Mum piped up, "Anyone here support Liverpool?"

"Mo Salah!" a young man yelled, and the T-shirt was duly handed over.

Our highlights of the trip so far had been defined by more than just the birds. It was 2018, the year of the FIFA World Cup, and we were going out of our way to watch every single England match. Wherever we went, every café, mountain lodge, and restaurant had soccer on. No matter how remote the location, the TV was blaring and everyone wanted to talk about the World Cup. Borders melted away in these discussions; you picked a side, and when they lost, you picked another, regardless of where you lived. (*Love Island*, too, preoccupied me more than it should have, and later in the trip, I made a special expedition into a local town to find out who had won.)

The campfire was now embers and I was tired, so I retreated to bed. The soft hooting of an owl was just lulling me to sleep when Dad leaped out of bed, followed by Mum.

"Mya!" they hissed loudly. "Get up."

Our guide was already at the door, about to knock. "Maybe eagle-owl!"

"Let's go!" Dad urged as I threw on my clothes. Hurriedly and by torchlight we followed our guide to the edge of the woodland that surrounded the lodge. Owls are easy to hear but notoriously difficult to see. And this might not be just *any* owl; it might be our target owl. Our guide held up a hand as we entered the trees. The owl hooted, our guide hooted—back and forth they went, calling and answering, while Dad shined the torch directly into the high branches. And then other lights appeared. A group of young lads were also aiming their lights into the trees. This was promising: the bird was rare enough to classify as an "event" even for the locals, so if they were here, we might be onto something.

And then we heard the quiet rustle of wings. The guide hooted and Dad's torchlight flashed. Soft wings flapped in the air, the amber eyes of the rare Usambara Eagle-owl watching us in the dark as it alighted on a low branch, its ghostly feline head turning to take in its audience. It sat for long minutes, heedless of the torchlight, our enraptured faces, and our silent awe.

As we approached the halfway mark in our trip, Mum's meds became an issue. The heat and length of the holiday meant we were already having problems keeping them cool. Until very recently, she had been enjoying herself; when she missed a bird, it wasn't such a big deal. This was a relief for two reasons. The first was that there was no way to get more meds if they went off; the second was the obvious one: if Mum was happy, so were we.

But now her dosage was becoming increasingly erratic, so much so that she sometimes hallucinated.

In hot countries, Mum prefers to sleep in arctic temperatures with lots of blankets, the AC on full power, sucking every bit of heat out of the room. Otherwise, she can't sleep. Dad and I are used to it. On the last night in our lodge in Usambara, Mum woke up just after midnight feeling too hot, convinced we would all suffocate in the heat. Half-asleep, she pulled open my mosquito net and yanked all the covers off my bed, and then she did the same with Dad's, after which she returned to bed. I'm a heavy sleeper, so her shuffling around didn't rouse me. But in the morning, I couldn't feel my toes. She hadn't fully closed my mosquito net, so my legs were covered in bites. Mum remembered what she'd done only when Dad and I started complaining. "It was hot!" she protested. "You should be thanking me."

We were in the Engikaret Lark Plains of Tanzania, famed, naturally, for their larks.

Moving through a monochromatic brown and scrubby landscape, the only things alive were the herds of cows and the Maasai shepherds who tended them. Although an utterly barren land, it is also the perfect habitat for the lark. The traditional deep red robes of our Maasai guides were the only bullets of color in this dusty terrain. We parked up and followed them into the scrubland while flocks of birds twisted in the wind.

The morning stretched on, and the ground began to bake. We walked on, looking into the hazy sky for any sign of larks. Hours passed in this way, and the heat began to fuddle my brain. What were we doing out here, stumbling around in blazing sunshine in the vain hope of spotting very small birds? I thought longingly of our icy bedroom.

"Over here!" hissed one guide, and I snapped to attention, turning around just in time to catch sight of a bird flying up from the parched earth and into the sky. It was a lark! But it had flown away before we had a chance to really inspect it; we spent yet another hour trying to track it down. Finally, on a patch of scrubby grass, we came upon a Beesley's Lark. *You're late!* it sang as it shuffled impatiently from foot to foot. This is a very rare bird, with perhaps one hundred individuals spread over its fifty-square-kilometer range. This isolated population had once been assumed to be a race of the Spike-heeled Lark found two thousand kilometers to the south, but recent studies have shown it is a species in its own right. With the arid microhabitat it now inhabits, it was isolated from the rest of the range of the Spike-heeled Lark five million years ago by the rise of Mounts Kilimanjaro and Meru and the subsequent change in weather patterns. They are an extremely difficult bird to see, partly because they are very adept at seeing you before you see them, and also because their habitat is so remote. They are an "event" bird, their sighting a cause for celebration among bird-watchers and non-birdwatchers alike. I wanted to stroke its peachy breast of soft feathers, a texture so incongruous in this desiccated environment it had to be touched to be believed. But, of course, I stood stock-still, like all good birdwatchers, and just stared.

Dad had taught me how to "digi-scope" during the Big Year—a method of taking a photo using your iPhone and a telescope—but I was yet to get the hang of it. We are not, in any case, master photographers, and the heat haze and the fingerprint smears on the camera lenses didn't help. But we did take a series of pretty terrible "record shots," just enough to prove we had seen the Beesley's Lark.

Madagascar separated from the African mainland more than one hundred million years ago, and most of its flora and fauna is endemic. For this reason, it is often referred to as "the eighth continent."

In the scorching midday heat, in the sandy savanna of Ankarafantsika National Park, in the middle of nowhere, I had connected with Ayesha to get my exam results, but it was the Malagasy Harrier that stole the show.

From the desert, we drove to the sea—to the Masoala National Park on the northeast peninsula of the island. I had just received an email from Chris Packham, inviting me to become his Minister for Diversity in Nature and Conservation and write an essay for his strident call to arms, the *People's Manifesto for Wildlife*. Eighteen strong independent voices (including Robert Macfarlane and George Monbiot), our "parliament of ministers" would highlight the most critical challenges to the UK landscape and its species. It was a great idea, a solution-led manifesto, and I was keen to take part.

The rainforest on the edge of the beach was a rowdy place: the songs of birds and other animals boomed during the day, and at night, the howls of lemurs and the chattering of insects filled the air. While Mum's rat phobia kept Dad on his toes, I was terrified of the large spiders, which thought nothing of perching on my pillow just as I was about to get my head down. Dad's job before he was allowed to go to bed was scouting every inch of the lodge for rats, bats, and spiders before Mum and I would even go into the room.

With hindsight, this was a bad idea—the more Dad searched for spiders, the more he found. The walls didn't properly join up with the floor, so for every spider he ejected, more followed. "Just get over it!" he would shout when he'd had enough. There was a small mezzanine floor in the lodge where the spiders were even more plentiful; that's where Dad slept while Mum and I huddled together in one bed downstairs, safe beneath a double mosquito net tucked in so tightly that no rodent or spider could enter. Dad probably welcomed the break.

I was, however, in the perfect environment to write an essay *about* the environment. I discussed the lack of diversity within the nature sector and how this situation has led to the sector's spectacular failure to attract VME people into nature. To have a real shot at saving our world, we must involve every ethnicity as a matter of urgency. Later in the year, Chris Packham organized the People's Walk for Wildlife. In front of a crowd of ten thousand people, young naturalist Dara McAnulty read a poem and I gave a speech, before we all marched to Downing Street to raise awareness of the catastrophic plight of wildlife in the UK.

. . .

The surrounding forest provided our only realistic opportunity of connecting with our target quest for the last two vanga species of the trip. As we were trekking through the open, undulating rainforest on the first afternoon, the Helmet Vanga, with its oversized, luminous, bright-blue hooked bill and rufous-and-black plumage, proved to be quite easy to see. First a single bird, then a pair, then a group of six picking around for large insects among the leaves and branches of the ancient trees. The rarer and smaller Bernier's Vanga was trickier to spot. We eventually spied a pair in the company of a feeding flock of their larger cousins, the rufescent female with her fine black barring outshining the all-black male. Pausing for a moment, they levered rotten bark and moss from the trees in their endless quest for food, and then they were gone. Both species are endangered by the twin threats to their habitat: slash-and-burn cultivation and climate change. Computer modeling predicts their ecological niche will have gone completely by 2050.

We had been traveling for almost ten weeks; each of us was showing signs of fatigue. Respite came in the form of torrential rain, forcing us inside for a couple of hours every day. Of course, this was ideal weather for spiders (and rats) to seek shelter, so Dad was kept busy.

When we arrived in Madagascar, Mum's spirits had been high, not manic, but now, she was beginning to unravel a little. Some of her meds had started to give off a funny smell, and we had had to throw them away.

Thankfully, we were nearing the end of the trip, and what better finale than a truly special bird? One afternoon, a couple of days before we were to head to the capital to fly back to the UK, Dad and I left Mum in the lodge to clear her email in-box while we ventured into the rainforest. This wasn't the dense jungle of Indonesia—the woodland felt light, airy, bright, and cool after the rains. I was hopeful we would see one more prize bird before we had to go home.

"Up there," Dad whispered, nudging me and pointing into the trees. My heart seemed to stop in my chest; for a moment I thought I was staring at the magnificent Harpy Eagle, one of my "top birds" for Ecuador when I was eight years old, which had, despite its prominent position on my short but expertly crafted list, evaded us—twice. We're not in South America, I reminded myself. Of course it isn't the Harpy Eagle. Instead, we were looking at an extremely rare Madagascan Serpent Eagle. Believed to be extinct, it was "rediscovered" in 1993. A medium-sized raptor, the Madagascan Serpent Eagle was a powerful presence all the same. A velvety dark gray and brown plumage adorned its back and wings, while its underbelly, neck, and thighs were a grid of black and white stripes. Its sharp talons sank into the branch on which it perched as its yellow eyes fixed us in a staring contest.

"Wow," whispered Dad. "Just bloody wow." We exchanged glances, silently acknowledging that one of us was going to have to leave the forest any second.

"I'll go," said Dad. "Unless you want to."

"That's okay," I said, returning my gaze to the eagle. "You'll be faster."

"Will I, though? I mean, you're younger..." he began, but I gave him a little push and he sprinted away, out of the forest, onto the beach, and back to our lodge, where he would find Mum and hurry her back to this very spot. If she missed this bird, there would be hell to pay—the Craig Family Harmony Index would collapse, perhaps forever.

When Mum and Dad returned fifteen minutes later, the eagle was still there and Mum was beside herself. Dad and I were just relieved the bird had stayed put. The unwritten rule among birdwatchers is that you always try to share your sightings, even if one of the party has chosen to stay behind to catch up on their emails. Mum's wrath, and indeed the wrath of any birder, would have been perfectly justified if we hadn't made the effort.

Our trip had ended on a high; we were ready to go home, ready for the next stage of the holiday, when we would pore over our lists, share our photos with friends and twitchers alike, and relive the most excellent highlights of a fantastic birding adventure.

Around this time, a new and compelling voice entered the climate change conversation. Greta Thunberg, a year younger than me, had started protesting outside the Swedish parliament holding a simple sign, SCHOOL STRIKE FOR CLIMATE. Greta's direct approach began to resonate with young people around the world, and by 2019, millions of students, me among them, had joined her rallying cry, skipping school on Fridays to demonstrate.

A year later, in February 2020, I shared a stage with Greta at the Bristol Youth Strike 4 Climate on College Green in the city center. In front of thousands of students and adults, I spoke about global climate justice. During an electric rally, charged with a desperate desire for change, I talked about the bigger picture, the hidden agenda behind climate change discourse, global climate justice. I posed the example of fast fashion to highlight the dilemma. When we in the global north are advised to stop buying the cheap clothes made in Bangladesh in order to reduce our carbon, water, and waste footprint, we must consider that Bangladesh is one of the countries least responsible for causing climate change. What will happen to those workers when we turn our backs on their products? Shouldn't we, the prosperous global north, given that our actions are the *most* responsible for climate change, consider supporting workers such as these to repurpose their factories with initiatives that would help the makers diversify their skills and move away from low pay and terrible conditions?

I was consolidating the future of my campaigning work on a big stage; it felt like a very important moment.

When Mum is manic, she latches on to one idea, and since we had returned from Africa, that idea was *me*. Endlessly productive and extremely proud of me, Mum was more like my manager than my mother, I had joked.

Ever since I set up Black2Nature, I had become a more vocal campaigner for inclusion, while remaining a very keen birdwatcher and climate change activist. People seemed interested in what I had to say, and I was regularly invited to give talks at nature conferences, as well as the odd appearance on *Springwatch* or *Countryfile*. Local and national press approached me for articles on either subject—or both—and interview requests poured in.

Mum was in charge of organizing my calendar, and she regularly slotted in events without checking whether I had the time, inclination, or energy to take them on. But she wasn't *depressed*, and probably for that reason alone, I went along with everything that appeared in my planner. Ever since I was a small child, Mum had instilled in me the idea that there was nothing I couldn't do so long as I had the drive to give it a solid go. But she also believed that if I stopped enjoying myself, I should question whether, in fact, it was time to quit. I wasn't enjoying my workload, and Mum had missed the signals that I was struggling. My desire to please just made the situation more stressful than it needed to be.

The nights were drawing in, and by the time I got home from school it was too dark to wander very far into the countryside, never mind go birdwatching. Like everyone else, I enjoyed the occasional weekend with nothing to do, and this particular upcoming weekend I wanted to go for a walk on the Mendip Hills, on my own. I looked at my calendar app, and immediately felt overwhelmed. Mum had inserted a radio interview on the weekend, without asking me. I felt like crying.

My dance class was in an hour, and I had been looking forward to stretching my limbs after days of sitting hunched over a desk. Right now, it was the last thing I wanted to do, but I also knew I needed the exercise. I slung my ballet pumps and leotard into a bag and raced downstairs. Mum was supposed to be giving me a lift, but she was no-where to be seen.

"Mum!" I screamed.

"I'm coming," she called, making her way down the stairs while draining a mug of tea.

"I'm going to be late!"

Once we were in the car, I let rip. "I don't want to ever do anything on a Saturday!" This wasn't strictly true—I had given interviews on the weekend before—but I wasn't in the right mood to be reasoned with.

"But they're expecting you," Mum said patiently. "You're tired right now, and it's days away, in any case. You might feel differently by then. Also," she added, "you did agree."

Had I? And then it came back to me, a quick exchange with Mum a couple of mornings ago as I raced out the door. I had forgotten.

By this point, I realized we were going to be late for my class, and I had been late, because of Mum, six weeks in a row. She was never quite ready when I was, and if that wasn't bad enough, she was often the last parent to pick me up. Time doesn't work for her the way it does for Dad, who is never less than punctual—at times even irritatingly early. By now I was furious and had started crying, insisting she turn the car around and take me home.

"Just splash your face with cold water before you go in," Mum said, turning into the car park of the local community center where the classes were held. "You need the exercise."

I jumped out of the car, hating that she was right, and headed straight to the loo. I duly splashed my face and then did my breathing exercises. I wasn't going to have a panic attack. Seven counts in through the nose, seven counts out through the mouth. Ten times. My heart rate slowed.

The lively piano music and repetitive exercises were soothing, and I began to feel better, but when I caught a glimpse of myself in the dark window, my resolve crumbled. My face was red and sweaty. I hadn't tied up my hair, and it hung in lank ropes around my shoulders. I looked miserable. I felt my throat close up, but I didn't want to cry in here, so I left the class to sort myself out back in the loo; maybe I would just wait for Mum to come and pick me up.

I didn't notice that someone had followed me out of the room.

When I was fourteen, I had been invited to give a TED Talk on the theme of "Passion, Priorities and Perseverance," in which I spoke about realizing one's goals, namely *my* goals, which were and always will be to watch birds and to encourage participation in nature regardless of a person's skin color.

In the cold bathrooms, I felt a warm arm drape itself across my shoulders and squeeze. Liv was in my class at school, an extremely cool and very popular girl. I was touched that she had abandoned ballet to check up on me. I guess she realized the hug wasn't making me feel better, so she began to chat. Liv told me that she had seen my TED Talk and had felt inspired. As usual, I was mortified that someone from school had watched me "in action"; I made a joke of it, mocking myself for how awkward I must have seemed. But Liv didn't like that—the talk had moved her, and she thought I should be proud of myself.

When she told me she had "stalked" me on YouTube and followed me on Twitter, I was even more embarrassed. Liv was acknowledging the public side of my life, the side that I had willfully chosen to believe was invisible to my friends. This was the point where, usually, I would try to shut down the conversation, but I was crying again and couldn't manage more than "Uh-huh." Liv, of course, didn't have a clue about my insecurities, she wasn't trying to give my ego a boost, and she wasn't being nosy either. She wanted, very simply, to cheer me up.

As she talked, something began to shift in my head, very slowly, like stiff gears on an old bike after they've been oiled. It wasn't an immediate transformation, but it was a start, a small thing that made a huge difference. This exchange launched me on a journey toward liking myself and being more wary of compartmentalizing my life. Liv remained supportive during the whole of sixth form. In her quiet, unassuming way, whenever she saw me in the corridors of school or out and about, she would mention she had read an article of mine or heard me on the radio or liked a tweet, and for a few minutes we would chat.

It began to dawn on me that it was time to leave my nest of insecurity.

I *was* Birdgirl: I had chosen the name, set up the blog, and taken on the work. Very gradually, I began to open up to a wider circle of friends; it wasn't so terrifying in the end because we are all much more interested in our own lives than in the lives of others. My media presence, campaigning, and avid birdwatching were one side of my life; my friends knew and liked me already and weren't fazed by the fact I had begun to acknowledge the other side. They were all aware of it anyway. All that had changed was that I was finally growing more comfortable in my own skin.

That Saturday, I did the radio interview and still managed a hike on the hills—with Mum. Binoculars to the skies, we marveled at the starling murmurations, shape-shifting clouds against a darkening sky. As we walked, I told her that I was happy to talk to whomever would listen about birds, diversity in nature recreation, and the environment, but to carry on talking, I needed to slow the pace and make more space for days like this. Striding over the hills, dusk settling all around us, Mum explained that the world was an amazing place; she wanted me to enjoy everything it had to offer. If there was a reason for her occasionally blustering approach, this was it: that I shouldn't miss any opportunity to do more, learn more, say more.

Was there a little sadness in her voice? Had she been denied such an active role in "life" because of her illness? She didn't say so, and I didn't ask—the answer would have been very hard to hear anyway. We carried on walking, raising our binoculars now and then as birds hove in and out of view, homeward bound, to roost for the night.

"I hear you, Mya," she said as we turned into the dead-end lane that would take us home.

I still feel awkward, still embarrassed, when people call out, "Hey, Birdgirl!" in the street or in the corridors of school, but these days I understand that I can't be two people in one body. I hadn't particularly enjoyed being a teenager: for most of that time, I had been stuck inside my own head, worrying about what others thought of me, while also navigating Mum's illness and Dad's anxiety.

Mum's bipolar disorder will always be a source of stress in our lives; her illness is a part of who she is, and each time she's depressed or manic, we're plunged right back into the familiar despair and fear that she might never emerge.

But when the three of us are twitching a rare bird, whether it's in the Hebrides, Madagascar, or the Somerset Levels, the Craig Family Harmony Index comes into its own, and I wouldn't choose to be with anyone or be anywhere else on earth. The bond between us has always been strong, and it's even more powerful when we're focused on our target bird, each of us knowing, without having to utter a word, that we're sharing, relishing, and celebrating a very special moment.

Epilogue

HARPY EAGLE

The Harpy Eagle is not only the largest eagle within its Neotropical home but is also among the largest eagles in the world. The female can weigh up to two times the weight of the male. With formidably sized talons, the Harpy Eagle catches (and feasts on) sloths and monkeys in the rainforest it inhabits. Even during the day, little light penetrates below the dense canopy of its forest home, and Harpy Eagles rely not only on their excellent vision to locate prey but on their keen sense of hearing as well. They have a facial disk of feathers, which can be raised to help direct sound to their ears—a trait they share with many owl species.

I long ago abandoned the notion of a life without storms, or a world without dry and killing seasons.

—KAY REDFIELD JAMISON, *An Unquiet Mind* (1995)

These days I am proud to call myself Birdgirl, but it was a winding road to acceptance. At the age of nineteen, I am happy to talk about birdwatching, traveling, diversity, climate change, and mental illness with whomever wants to listen.

If I had to choose one bird as my mascot, it would be the Harpy Eagle: fearsome raptors, named for the harpies of Greek mythology (part woman, part bird, and utterly terrifying). An odd choice for a mascot, I agree, but this magnificent bird is also a caring parent that has to fight hard for its survival. My family has fought hard for our survival.

I didn't see the Harpy Eagle when I went to Ecuador at the age of eight. It had been one of my "top birds," and yet it eluded me. I didn't see it on our second trip to South America, either, but in the summer of 2019, before the whole world shut down during the COVID-19 pandemic, I went to Brazil, and my dream of seeing this bird came true.

We were at the end of a superb five-week trip, first exploring the Atlantic rainforests of the southeast and then the Amazon. I had seen an astonishing number of new birds, over 350 in total, but—and it was a big *but*—there still was one itch I had not been able to scratch. Was this to be my third *unlucky* trip to South America?

An active nest site gives by far the best opportunity to catch up with this magnificent bird, but there had been none on our route, not even a couple of hundred kilometers *off* our route—that's how far we had been prepared to go. Between the laying of the initial egg to the final departure of a single young bird, the Harpy Eagle may attend its nest for upward of a year. This is great news if you time your visit well, but the adults only breed every two or three years, and that summer—rather thoughtlessly, I felt—they were between breeding cycles.

The previous day had seen us visit Bosque da Ciência, a research station located just outside Manaus, the capital city of the Brazilian state of Amazonas. What we didn't know was that there was a Harpy Eagle nest less than half an hour's walk from the center itself. That was the good news. The bad news was that the young bird had fledged and left the nest over six months ago; therefore, it was no longer active. Once more, we were in needle-in-a-haystack territory. The bird could be anywhere in this 11,000-acre site. Checking in with the researchers at the station, we discovered the eagle had last been seen five weeks earlier, when it made a quick stopover at the nest. Buoyed by the remote possibility it might come back, our guide led us into the rainforest, along a muddy track, which was more a flowing stream masquerading as a path. Twenty minutes in, and we were still squelching our way through the trees.

As we approached the nest, our guide asked us to wait while he went to investigate. Was this the moment I had been waiting for? How would I feel? Would it be an anticlimax?

"No sign of the bird," said our guide on his return.

It was a sobering moment, and I felt the familiar disappointment.

"Can we see the nest, at least?" I asked.

It was an enormous tree, tall and also very wide. Later, it would take the three of us to wind our arms around the trunk to hold hands. The nest, sitting atop an equally sizable bough, was huge—really staggeringly huge, easily big enough to seat a fully grown human with room to spare.

But what was *that*?

I was looking at the macabre sight of a skull and bones hanging from netting encircling the base of the tree: a sloth's skeleton. A recent happy harpy meal, no doubt. It was that or a monkey, another mammal this bird of prey is partial to. Here was some real evidence the eagle had been here, but it didn't appear, and eventually we squelched our way out of the rainforest.

We were returning to the UK the following day, so I had one more shot. The next morning, my boots still sodden, we set out for one final attempt. It felt like a familiar route by now: get out of the vehicle, check in at the research center, walk down the main track, cut off along a path to the left, walk through the mud and water. Could my boots get any wetter? Yes, it appeared they could. Our guide left us alone once more to recce the site. I wasn't hopeful, and my mind drifted to other things: our imminent return home, the start of my next school year—

I was snapped out of my reverie and back into the rainforest by a fierce whisper through the trees. "Come! Now!" hissed our guide.

Before I knew what was happening, Dad was propelling me along the path. No doubt Mum was pushing him—we were going fast and slow at the same time, keen not to trip, but we didn't want to miss this moment. Who knew if the eagle would wait for us?

In front of the towering tree, I let out a gasp. There was a juvenile Harpy Eagle, perched on the branch above its nest.

How to describe this moment? Elation, surprise, relief, disbelief— emotions coming thick and fast. Tension dropped away, and excitement kicked in. I slowed my breathing and focused on the bird, drinking it in. For nine years I had longed to see this fine animal, and now here it was—or rather, here *she* was. Far bigger than the male, she inspired a thrill of fear in me as I stared up at her. She was eyeing me just as closely. Did we look like monkey morsels from her towering height?

The Harpy Eagle is one of the largest eagles in the world. Their eyesight is eight times as powerful as a human's; their talons are larger than a grizzly bear's claws and more powerful than a rottweiler's jaws. The juvenile's plumage was white, with a gray breast band, but it was her very strange, very menacing face that was most intriguing. She looked like a cross between an owl and an eagle. A crest of snowy feathers lay flat against her head—when the bird is agitated, its crest stands on end to form a fearsome crown.

Maybe I should have been scared when her feathers flicked up, but all I felt was an overwhelming, childlike excitement. I wanted to jump up and down, punch the air, and scream, but I am an obedient birdwatcher, so I didn't speak or move. Soon I was barely breathing. I forgot about my wet feet, the mud, and everything else as my focus narrowed to the fantastical creature in front of my eyes.

For long minutes we checked each other out. When she had had her fill of us, she spread her wings and heaved her giant body into the air.

Harpy Eagle? Tick. Over five thousand birds seen to date? Tick. Brazil had brought me to this landmark magic number: I had seen half the birds of the world.

In 2020, while I was preparing for my A levels, the COVID-19 pandemic was sweeping the world. One day, at the end of March, we were all sent home from school. I was never to return. I went from looking forward to a gap year, during which I'd planned a trip by train to the United Nations Convention on Biological Diversity in China, following the Trans–Mongolian Railway, to sitting around at home with my parents *for months*.

In August, I received an email from Greenpeace, inviting me on their science expedition to the Arctic. Was I interested in the role of social media spokesperson to highlight their Thirty by Thirty campaign? Yes, of course I was. The mission was to secure the protection of 30 percent of the oceans by 2030. As the Arctic ice melts, new seas are revealed; it was Greenpeace's ambition to gather evidence that life exists in these new waters and therefore should qualify for protection.

I had spent an increasing amount of time over the previous six months campaigning against climate change and for global equality within the movement, to give those in the global south a voice. I spent lots of time speaking on Zoom, streaming on Instagram Live and YouTube Live, giving interviews, writing articles, and using whatever platform I had access to in order to talk about climate change, and specifically about global climate justice.

I had previously given a talk to Greenpeace employees on this topic in an attempt to fight ignorance within the environmental sector. Challenging campaigners to think about their preconceptions was hard work. This area of work, along with loss of biodiversity, has come to define my activism, given I have firsthand experience of each from my travels.

Pandemic restrictions meant ten days of quarantining in Germany before we embarked on our journey to the Arctic. I was in Cuxhaven when Greta Thunberg announced an international day of action for climate change. My first thoughts were of Bristol, how I would miss protesting with my friends. But very quickly, I realized I was going to be in exactly the right place.

I wasn't prepared for the scenes that greeted me when we arrived in the High Arctic after five days at sea: everything was melting, ice sheets floated freely and away from us on our approach.

During our visit, we were never to reach the main ice sheet, and most of the time it felt as if we were floating in a giant bowl of slush. It was devastating and strange to imagine that the fast-melting ice had the power to affect the lives of people on the other side of the world, in places just like Bangladesh, where there was terrible flooding. Even a small increase in temperature has the potential to speed up global warming. I thought about just how interconnected everything was, and my unhappiness gave way to anger.

On the morning of September 25, the Global Day of Climate Action, I was giving an interview to a Reuters journalist about the impact of the declining ice at the poles, when a bird caught my eye. I had come to the Arctic armed with a target list, of course, and one of these was the Little Auk. "Give me a minute?" I asked, raising binoculars to my eyes. Little Auks are not rare birds, and most experienced birders have seen them, but somehow, just like the Harpy Eagle, they had always eluded me. And now, here it was—not much bigger than a starling, black all over apart from an appropriately snowy white belly—its wings beating furiously as it flew past.

Later that morning, in a dozen layers of clothing, the ice melting around me, I positioned myself on an ice floe, holding a simple sign that read YOUTH STRIKE FOR CLIMATE. The Reuters team took photos and posted them to global news platforms. Within minutes, my phone started pinging: the picture had gone viral. The image appeared on the front pages of newspapers and on TV channels around the world, off-line and online, reaching remote places and people with a global message on climate change.

I wasn't too surprised; having been involved in campaigning for years, I knew that a dramatic tableau was more likely to make an impact than a carefully worded blog post. It felt like a defining moment. I was out there for five hours, holding up a simple sign while I hopped between icy islands.

Before we returned to Germany, I swam in the glacial waters of the Arctic and recalled my Antarctic swim. Would it be as cold? Yes. It was.

I had left the UK just as we were easing out of lockdown; by the time I returned, we were back in the thick of it.

As I approach my twenties, I am filled with anticipation for the future, filled with hope, too, that Mum will enjoy longer stretches of peace, that Dad will find the space he needs to enjoy the birds, the hills, the fresh air, and travel.

University beckons and a new stage of my life will begin. Black2Nature remains my biggest priority, to make sure the foundations are strong enough for our message to thrive as I take a small step back. I created the platform as a young teenager, and now it is an independent charity, with its own momentum and voice. Our message is as powerful as ever: We demand equal access to natural spaces for minority ethnic communities. We will continue to engage the VME sector of the community, significantly less engaged with nature, with climate change and environmental issues. If 20 percent of the population comes from a VME background, we have little chance of fully embracing our environmental challenges if this 20 percent is excluded. We will carry on talking, writing, and meeting with the nature sector to highlight the racism that must change if the natural world is to become a truly egalitarian space populated with vocal, caring, and driven individuals.

The message on climate change is a simple one: cut back on carbon production. While this is certainly needed, it is, at the same time, far too narrow an approach. Call me a disruptor, I don't mind; my aim has always been to challenge lazy beliefs, where dire consequences haven't been considered. I plan to carry on campaigning for global climate justice. The intention to save a habitat is one thing, but how this intention is executed is another. In the Democratic Republic of the Congo, for example, the World Wildlife Fund (WWF) funded and trained rangers who committed human rights abuses against the Baka tribespeople in their efforts to form a national park, evicting people from their land, beating and imprisoning them when they resisted. I reassessed my views on poaching when I discovered allegations that in Cameroon and Nepal, the WWF funded paramilitary guards to challenge the poaching practices of local people. These "poachers" had lived and hunted on what was once their land for generations, before it had been "reclaimed" as part of a conservation project.

Human rights abuses suffered by Indigenous peoples kicked off their land in the name of conservation are racist. I'm campaigning for a "just transition," equal opportunities, and support for the global south during the period of radical change that must be made to save our world. As an ambassador for Survival International, I will continue to work as a spokesperson in behalf of the human rights of Indigenous peoples.

It goes without saying that I appreciate the experiences I have had. I have traveled to forty countries and seven continents, seen over five thousand birds, including my favorite, the Harpy Eagle. I would still give them all up in a single second, however, for Mum to be well; for Dad to have more time to do what he wants to do; and for Ayesha, now a single mother of two, to have a more balanced life. But Mum is still very ill, and my dad veers between exhaustion and sometimes depression.

Thinking about these trips, two words come up again and again: *respite* and *regeneration*. Our trips fed our obsession, and we were regularly lost in a shared hunger to see birds: rare birds, the endemics, glorious birds of paradise, and even squat brown ones who prefer the undergrowth to the treetops. Whether we were sweating in dry savannas or moist rainforests, going on mountain hikes or camping down in the snow, we were *outside*, under the sun, or the moon, breathing in unfiltered fresh air, on a mission. How could these trips be anything less than a respite from life at home as well as a period of regeneration?

We went away for three weeks or six weeks or six months, and while we were still essentially ourselves, we had been taken out of context. Mum wasn't confined to her study, red eyes glued to a screen, and Dad wasn't waiting for her to come to bed or watching her like a hawk while she swallowed her meds every morning and night. And I could abandon, for a while anyway, the dual narrative that defined so much of my teenage years.

The trips were certainly a period of healing, bonding, but their value extended far beyond the time we were actually away. We all felt lighter in the weeks leading up to a trip, as we drew up our target lists, researched habitats, booked guides, and studied routes across vast countries. And afterward, we lived on our memories as we consolidated these lists, posted our blogs, and shared our birding experiences with the community—and in this way, they sustained us over the winter months.

Mental illness motivated us to travel, but it was being together that kept us on the road. Mum thrives when we're in defined spaces, whether that's a car, van, tent, bedroom, or open country. She relishes spending every minute of every day within the bubble of our family for the period of time in which we're away. It doesn't matter if she and Dad are screaming at each other about directions or full campsites—she is distracted and entirely in her element.

"A life without storms...."? My family's story isn't one that culminates with an inspirational "happy ever after." We have shared so many magical experiences, and have often gone for days without so much as a glimmer of mania from Mum, nights when we stayed up to watch owls fly across starry skies. But we always returned to ourselves, to the strain of living with a member of the family who might never truly get better and another who is often buckling under his load. These days, when Dad has had enough, I take over for a day or two while he finds a friend to walk the hills of another county with. I sleep in Mum's bed, make sure she takes her meds, and when they surface, I try to calm her night terrors. This is the way we get by, a day, a week, a month, a year at a time. But there will always be storms.

My father's role has been one of resilience and a desire to keep our family together. When my family was in crisis, Dad was often depressed but not despairing. He was never immobile, and his forward motion inspired us to keep moving. He identifies the hurdles in our lives and gets us over them. He has encouraged me to always prioritize experience over possessions and to remember that what we do in our spare time defines us. For Dad, it's a simple trade-off: he would choose to be on the hills with a pair of binoculars, where time is the greatest currency. For me, it's increasingly about campaigning for a fairer world. If I have learned one thing from him, it's that a situation is never so overwhelming as it might first appear. These days, I am better at standing back, taking a minute, breaking down the issue, and waiting until an answer presents itself. My voice is louder, more assured, as I strive to make well-prepared and well-executed contributions to the environmental agenda. I learned curiosity from my father, that there is always more to see, further to go, effort to exert if you are to achieve your goal.

Mum has taught me how to think, how to take apart ideas and keep adding to the conversation. She also believes that relationships are worth nurturing, and the best way to do this is by doing something you love—together. And that "something" has to be an activity that everyone is happy enough to take part in. In the Venn diagram of interests, it's the bit in the middle.

My life would not be what it is today if I didn't love birds. Mum and Dad wouldn't be together if we hadn't traveled to find birds. I wouldn't be campaigning for diversity in nature if, as a young child, I hadn't been exposed to such a homogeneous community, all while searching the skies for birds.

The birds' simple, instinctive lives have guided me over the years to listen, to watch, and to exercise patience. These principles feel like good enough rules to live by.

Acknowledgments

Firstly, and most of all, thank you to Mum and Dad, without whom I would not be the person I am today. For sharing so many amazing birding trips and experiences with me; for filling my head with ideas of social and environmental justice, politics, animal rights, and activism; for being my personal assistants, secretaries, and chauffeurs over the years; and for making me believe that, with hard work, I could achieve anything I put my mind to. For supporting me as a teenager when I wanted to organize nature camps, conferences, and then set up my charity, Black2Nature, and for feeding my obsession for music, books, and films. Also, to Dad, for enthusiastically teaching me everything I know about birding, and when, aged nine, I showed an interest in bird ringing, for dedicating his weekends to our learning together. To Mum, for teaching me all things Bangla, the joys of singing and dancing Bollywood style to *Pakeezah*, as well as how to fight for antiracism, equality, and equity.

To my older sister, Ayesha, for being the cool birding role model that most young women lack, for being a second mum as well as a sister, always being there for advice whether asked for or not, and for the countless hours committed to organizing and running my nature camps for visible minority ethnic young people over the last seven years.

To Laila and Lucas, because every child needs a nephew and niece to boss around, and for their enormous enthusiasm and help at camps!

To Gran and Nanu, the loving and supportive matriarchs of my family, and to Granddad and Nanabhai, both of whom died too young, before I could meet them. Their legacies live on in my love of birds and my passion for activism.

To my endless jumble of aunts, uncles, and cousins—a wonderful, messy, complex family, as all families should be. To my *boro khala*, Monira Ahmed Chowdhury, and *chuto khala*, Lily Khandker, who have dedicated their lives to fighting against racism and for the equality of minority ethnic and marginalized communities, and who inspired me to do the same. Also, to Auntie Penny, a dedicated children's nurse, who taught me selflessness and compassion; to my mamas, Hasan and Faisal, who educated me on topics such as "the history of empire" and "colonialism now"; and to all their spouses, my cousins, and other family members for turning up to cheer me on.

My book has been created through the love and care it has received from the "*Birdgirl* team." I could not have done it without you. Thank you to Claire Paterson Conrad, my literary agent, for seeing something in the eighteen-year-old me that I had not yet imagined, believing in my story, and guiding a naïve teenager and her parents through the world of publishing; to Arzu Tahsin, editor extraordinaire and now a lifelong family friend, for supporting me through the journey of turning my book proposal into an actual book, for helping me to shape and create *Birdgirl*, and for allowing me to enjoy the process and have fun; to Michal Shavit, publishing director at Jonathan Cape, for incisively seeing through the mist of words to the core of what I wanted to say, for believing in *Birdgirl*. And, of course, the rest of the team at

ACKNOWLEDGMENTS

Cape: Joe Pickering, Rosanna Boscawen, Oliver Grant, Alison Davies, Chloe Healy, Bethan Jones, Shabana Cho, Sarah Davison-Atkins, Bea Hemming, David Milner, Ana Fletcher, Ruth Waldram, Justin Ward Turner, Christina Usher, Daisy Watt, Cecile Pin, and Sasha Cox, along with my wonderful cover team: jacket designer Gabriele Wilson, artist Thomas Davies, and photographer Mack Breeden for taking my story and creating a thing of beauty, a book, ready to make its own journey into the world.

Thanks also to Chris Clemans, Roma Panganiban, and the team at Janklow & Nesbit Associates and Deb Futter, Randi Kramer, and the team at Celadon Books for their support and belief in the North American edition of *Birdgirl*.

To Josephine Thurley, Flora Webber, and Megan Jones at ITG, the collective powerhouse of the *Birdgirl* team, for their support and guidance well beyond the call of duty, and especially their WhatsApp prompts ("M I've put the zoom link into your diary." "M have you signed onto your meeting yet?" "M they're waiting!"); and to Preena Gadher and team, Caitlin Allen, Emily Souders, and Angel Pearce at Riot Communications for taking on and organizing a messy Gen Z who doesn't read emails or messages.

To the organizations and people who have entrusted me with positions of responsibility and supported and educated me, including Survival International, Greenpeace, the Wildlife Trusts, RSPB, Froglife, Beaver Trust, Burns Price Foundation, the Bristol Global Goals Centre, Creative UK, and the Summer Camps Trust.

Thank you to the legend that is Bill Oddie, for talking to a young girl at a book signing and making an immediate connection through our shared experiences of our mothers' mental health issues; for agreeing to be the keynote speaker at my first Race Equality in Nature conference back in 2016 when I was fourteen, when few people had heard of me, and for being there over the years when I needed someone to talk to about the impact of bipolar disorder. To Chris Packham, for believing in and supporting me from the start, trusting me to speak in front of ten thousand people in Hyde Park, and appreciating my "punk rock attitude!"; to Liz Bonnin, for continuous, quiet support, empathy, and understanding over the years. To Steve Backshall, for *Deadly 60*, which I watched endlessly as a child, providing me with the passion to seek out wildlife around the world. And, of course, to David Attenborough, whom I have had the pleasure of meeting, for inspiring so many people over so many years to love and respect nature.

To Emma Watson, for amplifying my voice.

The two Richards, Benwell and Pancost, for championing my views and sharing their own with me over the years, helping my thoughts develop and grow. And to Rich Pancost, for supporting my "spoof" honorary doctorate award from the University of Bristol—which turned out to be real!

Thank you to Mike Bailey, my bird ringing trainer, for passionately teaching me the science behind ringing, and all those at Chew Valley Ringing Station. To the finders of rare birds in the UK, for sharing the joy of your finds; and the wonderful bird guides around the world, for sharing your countries' birds, history, stories, and experiences of Indigenous peoples and culture. Andrés Vasquez, the birds were "quite nice!"

To my friends, thank you for all your love and support.

And, finally, to all activists everywhere with their own untold stories: keep believing in and fighting for a better world.

About the Author

MYA-ROSE CRAIG, also known as Birdgirl, is a twenty-year-old British Bangladeshi birder, environmentalist, and diversity activist. She campaigns for equal access to nature and to end the climate and biodiversity loss crises, issues that she believes are intrinsically linked, while promoting global climate justice. She also fights for the rights of Indigenous peoples, is an ambassador for Survival International, and has previously written a book in the UK amplifying Indigenous voices.

At age fourteen she founded Black2Nature to engage teenagers of color with nature, and at seventeen she became the youngest Briton to receive an honorary doctorate, awarded by the University of Bristol, for this pioneering work. Also at seventeen, she became the youngest person to see half the world's bird species and shared a stage with Greta Thunberg, speaking to forty thousand protestors. In September 2020 she held the world's most northerly Youth Strike for Climate, traveling with Greenpeace, for whom she is an Oceans Ambassador, to the melting pack ice of the High Arctic.

Founded in 2017, Celadon Books, a division of
Macmillan Publishers, publishes a highly curated list
of twenty to twenty-five new titles a year. The list of
both fiction and nonfiction is eclectic and focuses
on publishing commercial and literary books and
discovering and nurturing talent.